Lean Six Sigma Green Belt Slide Book

Lean Six Sigma Green Belt Slide Book

Mary McShane-Vaughn

University Training Partners Publishing

University Training Partners Publishing

Folsom, CA

https://courses.6sigma.university

Library of Congress Cataloging-in-Publication Data

McShane-Vaughn, Mary

    Lean six sigma green belt slide book/Mary McShane-Vaughn

       p.cm.

ISBN-13: 978-0990683872 (University Training Partners)

1.   Quality Control – Statistical methods – Handbooks, manuals, etc.

    I. McShane-Vaughn, Mary II. Title

    First printing 2020

Introduction

Congratulations on starting your Lean Six Sigma journey! This course manual includes a study planner, all the slides from the module videos, and the case study used in the Lean Six Sigma Green Belt course from University Training Partners. You will refer to this text throughout the course as you complete assignments and take the module quizzes. Feel free to annotate and tab the text as you go. The course manual will also prove to great takeaway and reference as you start on your first Lean Six Sigma improvement project. Enjoy!

To learn more about other Six Sigma and statistics courses University Training Partners has to offer, please visit our website: https://courses.6sigma.university

Regards,

Dr. MMV

Mary McShane Vaughn, Ph.D.

Founder

University Training Partners

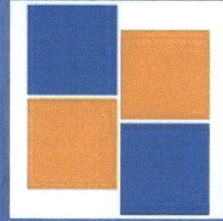

Study Planner

Lean Six Sigma Green Belt

Welcome!

Congratulations on starting on your Lean Six Sigma Green Belt!

A Green Belt is an important, and marketable, credential to have. Companies are looking for employees who can select the right improvement projects, manage a project team, make data driven decisions, and implement real improvements in processes. Improvements that translate into better bottom line results (i.e. $$$). As a Lean Six Sigma Green Belt, you will be able to do all of that.

This course will change the way you approach projects, the way you make decisions, and the way you see your role in your organization. Along the way, you'll have some "aha!" moments as you:

1. Get very familiar with the Six Sigma project cycle

Six Sigma concentrates on improving processes by initiating projects that follow five distinct steps: Define, Measure, Improve, Analyze and Control. These steps will roll off your tongue by the end of the course, trust me.

2. Change the way decisions are made in your organization

The decisions Green Belts make with their teams are driven by data, not by whoever has the loudest voice in the room, or the highest seniority, or the most intimidating personality. Think for a minute about your workplace: how are decisions usually made? What are the driving forces behind them?

3. Learn the right way to measure project outcomes

The goal of Six Sigma is to decrease variation – to be more consistent with our process inputs and outputs – and to decrease defects. Once we do this, we also increase customer satisfaction and improve bottom line results.

Are you excited to begin, or are you a bit scared of the math?

Sure, there is some math involved with this course. We're making decisions based on data, so there has to be some analysis, right? If you are comfortable with creating graphs and using formulas in Excel, you're way ahead of the game. If you are not, you'll be adding these tools to your already overflowing Green Belt toolbox. There are quick tutorials included in the course to help you along the way.

The truth is the hardest part of leading Six Sigma teams is not the data collection or statistical analysis – it's dealing with people. The "soft skills" of project selection, project prioritization, project scoping, stakeholder analysis, team building, and the voice of the customer are all employee-facing or customer-facing – and difficult. As I tell my students, you can always hire a contract statistician to do the heavy lifting for you if you run into a complex analysis situation, but you cannot outsource managing your people.

Feel better?

If you get stuck, you can always send me an email or post a question on the discussion board.

You've got this.

Now let's get started!

Course Design

Self-paced modules

This self-paced, online course consists of 14 separate modules that introduce you to Lean and Six Sigma and then systematically fill your toolbox with the most useful techniques for identifying and successfully running your first improvement project. You will have access to the information in these modules for 6 months.

Case study

Throughout the course, we'll be using a case study based on a resort hotel. This will allow you to practice applying the tools you learn in each module to a real-world situation.

Module design

For each module, there is an introduction that lets you know what you'll be learning. Then, there are a series of short videos that introduce the module material. There will be an assignment to submit, as well as a knowledge check for each module.

Assignments

There are short assignments for each module that will let you practice what you have learned. Many assignments will be based in Excel. Your instructor will review your submission and either accept it or turn it back to you for edits.

Don't feel bad if an assignment gets turned back to you. As Dr. MMV says, we learn the most from making mistakes. Consider a returned assignment an opportunity to grow and master a skill. Just make changes to the submission based on the instructor's detailed feedback and resubmit for another review.

To successfully complete the course, you'll need to have all assignments accepted by the instructor.

Knowledge checks

There is also a knowledge check for each module consisting of multiple-choice questions. You can make as many attempts as needed to score at least an 80% on each knowledge check. Once you meet the score requirement, you will be able to move on to the next module.

Instructor support

If you get stuck, or have a question or insight to share, there is a discussion board for each module. Dr. MMV monitors these boards and will answer any questions you may have. Just because this course is self-paced doesn't mean you are on your own. Reach out whenever you need to!

Course outcomes

When you have successfully completed the 14 modules, and have had all assignments accepted, you will receive a downloadable certificate showing that you have completed the Lean Six Sigma Green Belt course requirements. This is certainly a time to celebrate and brag a bit on your social media feeds!

Planning Your Time

You will have 6 months, or 180 days, to complete this course. It's best to put yourself on a schedule to assure that you maintain your momentum. We've designed a 14-week study plan for you to follow. Even if you get off track by a week or two (life does happen), you'll still have plenty of time to finish and earn that Green Belt credential!

Week	Module		Learning Activities
1	Welcome to the course!		Why learning to be a Lean Six Sigma Green Belt is important
			How this course is designed
			How to use this course
			Before we begin...
	Module 1: What is Lean Six Sigma?		What you'll learn in this module
			Lean Six Sigma introduction
			Lean Six Sigma introduction - part 2
			The DMAIC Roadmap
			Test your learning

Week	Module		Learning Activities
2	Module 2: The Mirasol case study	📄	What you'll learn in this module
		⬇	The Mirasol Hotel case study
		✏	Assignment: Case study response
		✏	Assignment: Force field analysis
3	Module 3: The voice of the customer	📄	What you'll learn in this module
		▷	Capturing the voice of the customer
		▷	Kano analysis
		✏	Assignment: Kano survey
		▷	Using Kano survey results
		⚙	Test your learning

Week	Module		Learning Activities
4	Module 4: Six Sigma projects		What you'll learn in this module
			Project selection
			The project charter
			Assignment: Mirasol project brainstorming
			Assignment: Project prioritization matrix
			Test your learning
5	Module 5: Leading people		What you'll learn in this module
			Stakeholder analysis
			Using the stakeholder template
			Six Sigma teams
			Assignment: Stakeholder analysis
			Test your learning

Week	Module		Learning Activities
6	Module 6: Processes and graphing tools		What you'll learn in this module
			The cost of quality
			Process maps
			Graphical tools
			Pareto chart how-to
			Run chart how-to
			Assignment: Pareto chart
			Assignment: Run chart
			Test your learning

Week	Module		Learning Activities
7	Module 7: More graphing tools		What you'll learn in this module
			Scatter plots and histograms
			Scatter plot patterns
			Scatter plot how-to
			Assignment: Scatter plots
			Test your learning
8	It's Ninja time!		Whew! You are halfway there
			Who's a ninja?
	Module 8: Descriptive statistics		What you'll learn in this module
			Descriptive statistics
			Formulas how-to
			Assignment: Descriptive statistics
			Test your learning

Week	Module		Learning Activities
9	Module 9: The normal distribution	📄	What you'll learn in this module
		▷	The normal distribution
		▷	Confidence intervals
		✏️	Assignment: Z scores
		✏️	Assignment: Confidence intervals
		⚙️	Test your learning
10	Module 10: The 1.5 sigma shift	📄	What you'll learn in this module
		▷	Sigma levels and the 1.5 sigma shift
		▷	Sigma level examples
		✏️	Assignment: Sigma levels
		⚙️	Test your learning

Week	Module		Learning Activities
11	Module 11: Capability analysis	📄	What you'll learn in this module
		▶	Process capability
		▶	Capability examples
		✏	Assignment: Calculating capability indices
		⚙	Test your learning
12	Module 12: Control charts	📄	What you'll learn in this module
		▶	Control charts
		▶	Which chart?
		▶	Chart patterns and out-of-control conditions
		▶	Control plans
		⚙	Test your learning

Week	Module		Learning Activities
13	Module 13: Lean Tools	📄	What you'll learn in this module
		▷	Visual management, 5S and mistake proofing
		▷	Eight wastes and SMED
		▷	Quick changeover: Formula One Pit Crew
		▷	Meals per hour
		▷	5S in the Office
		▷	5S in a hospital supply room
		⬇	Lean in the airline industry
		✏	Assignment: Eight Wastes
		⚙	Test your learning

Week	Module		Learning Activities
14	Module 14: Case wrap up	📄	What you'll learn in this module
		▷	Mirasol project reports – Jim
		▷	Mirasol project reports – Eli
	Next steps	📄	Congrats! Here's what's next…
		📋	Before you go…

Course Progress

Keep track of your learning by checking off your accepted assignments and knowledge checks for each module using this check list.

Module	Deliverable	Check
Welcome to the course!	Before we begin…	☐
Module 1: What is Lean Six Sigma?	Test your learning	☐
Module 2: The Mirasol Case Study	Assignment: Case study response	☐
	Assignment: Force field analysis	☐
Module 3: Customers	Assignment: Kano survey	☐
	Test your learning	☐
Module 4: Projects	Assignment: Mirasol project brainstorming	☐
	Assignment: Project prioritization matrix	☐
	Test your learning	☐

Module	Deliverable	Check
Module 5: People	Assignment: Stakeholder analysis	☐
	Test your learning	☐
Module 6: Processes	Assignment: Pareto chart	☐
	Assignment: Run chart	☐
	Test your learning	☐
Module 7: Graphs	Assignment: Scatter plots	☐
	Test your learning	☐
It's Ninja Time!	Who's a ninja?	☐
Module 8: Descriptive statistics	Assignment: Descriptive statistics	☐
	Test your learning	☐
Module 9: Normal distribution	Assignment: Z scores	☐
	Assignment: Confidence intervals	☐
	Test your learning	☐

Module	Deliverable	Check
Module 10: The 1.5 sigma shift	Assignment: Sigma levels	☐
	Test your learning	☐
Module 11: Capability analysis	Assignment: Calculating capability indices	☐
	Test your learning	☐
Module 12: Control charts	Test your learning	☐
Module 13: Lean Tools	Assignment: Eight Wastes	☐
	Test your learning	☐
Module 14: Case wrap up	N/A	N/A
Next steps	Before you go…	☐

MIRASOL RESORT

A Lean Six Sigma Case Study

JONATHAN SAND

President Sand Island Resorts

This case is based on a small beach resort hotel recently acquired by Jonathan Sand, President of Sand Island Resorts, LLP. Sand began working in the hotel industry as a teenager, manning the front desk and taking reservations at the local hotel owned by his uncle.

After earning his college degree in manufacturing management, he worked as an assembly line supervisor at an automotive plant as well as a production engineer in the pressed metal shop. When the assembly plant was shut down, Jonathan returned to the East Coast and found a position as a maintenance supervisor at Mirasol, a local beach resort. He found that his industrial background gave him a unique edge in attacking and solving problems in the hospitality sector.

In the past five years, he has held wide-ranging positions at the resort to include accounting supervisor, banquet manager, and housekeeping manager. In 2008, Sand was promoted to general manager of Mirasol, which was then owned by Allendale Holdings, a company that managed seven high-end resorts in South Carolina and Florida.

PAGE 2

VISIT SOUTH CAROLINA

Takeover from Allendale Holdings

The economic downturn and credit crunch forced Allendale to liquidate some of its properties at deeply discounted prices in order to remain solvent. When it was announced that Marisol was being sold off, Jonathan contacted his uncle, now retired, and formed the Sand Island Resorts partnership. Sand and his Uncle Fred were able to acquire the property a year ago. Since then, Sand has been enthusiastically making changes to put his own stamp on the resort.

In 2009, the resort was featured in the Condé Nast Traveler magazine, which described it as "a boutique resort that is the ultimate in a relaxing getaway; private and personalized."

"A BOUTIQUE RESORT THAT IS THE ULTIMATE IN A RELAXING GETAWAY; PRIVATE AND PERSONALIZED."

Inspired by the Condé Nast review, Sand crafted the new company mission statement and has it displayed in framed posters that hang in the main lobby and the back offices:

"Sand Island Resorts is committed to providing its member guests with personalized service and a relaxing resort experience that consistently delivers excellent quality and high value."

FACILITIES

The Mirasol resort is located on a small barrier island along the Intercostal Waterway, accessible only by passenger ferry. The ferry, operated by the State, makes six round trips each day. Automobiles are restricted on the island, and the main means of transportation are golf carts and bicycles.

The small resort has 50 suite-style rooms in the two-story main building. There are two pools, one with a swim-up bar, as well as three heated spas on the property. Water sports are accessible from the marina located to the west of the property, reachable by a complimentary golf cart shuttle. The beach is accessible via a boardwalk, and guests can reserve cabanas, umbrellas and chairs, surf boards, boogie boards and snorkel equipment at the beach pavilion.

There is one restaurant serving breakfast and dinner daily in the main building, a small bar located off of the main lobby, a poolside snack bar, and a separate banquet/meeting facility that can accommodate up to 175 guests. A gift shop is in the lobby, selling sundry items and resort-branded tote bags, beach towels, golf shirts and Tervis tumblers. A separate exercise and spa facility is located next to the pool area. The gym has four stationary bikes, three treadmills, a Stairmaster, an aerobics/yoga room and locker rooms with showers. The spa has two treatment rooms, a manicure and pedicure station and 2 outside massage cabanas.

The facilities building is located at the north end of the property, and contains a commercial laundry facility, a maintenance garage for the resort golf carts and bicycles, a tool shop, two storage rooms, a hurricane preparedness room and the engineering department offices.

ECONOMIC & MARKET PRESSURES

Allendale Holdings saw a 15% decline in room occupancy and special events bookings at the start of the economic downturn in 2007. Since purchasing the resort, Sand Island Resorts has experienced a 2-3% growth in these rates.

Sand attributes the growth to ads he ran in newspapers in the Northeast, and would like to expand the advertising campaign. His uncle would rather concentrate advertising efforts on adjoining states due to the cost of gasoline. Reina Andersen, Marketing Director, has proposed partnering with online booking sites to offer special accommodation packages at deeply discounted rates. Sand is reluctant to run the special rates because he is afraid of diluting the Mirasol brand.

In a recent meeting, the Director of Recreation and Spa Services, Linda Petrucelli, suggested offering a spa destination package during the "shoulder season," complete with spa cuisine, spa treatments and weeklong courses taught by celebrity yoga masters. Sand thinks that demand for such event is not strong enough to offset the cost.

While Allendale was experiencing a cash flow shortage, it postponed routine maintenance and upkeep. Upon acquiring the property, SIR had to perform costly repairs to the pool deck, air conditioning system and septic system.

In addition, SIR repainted the exterior, made improvements in the restaurant and spa facilities, and purchased new furniture for the beach cottages.

Over the past year, the cost of maintenance, repairs and improvements have eaten into SIR profits. Controller Greg Schultz has suggested reducing the number of resort amenities to offset increased expenses, but Sand is worried that fewer services will translate into fewer bookings.

In addition to economic worries, Sand is concerned that a new hotel recently opened on the south end of the island will dilute his market share. The hotel is an all-inclusive resort owned by a national chain that caters to families with children. He has a suspicion that that his more price sensitive customers are flocking to the new hotel because it offers more activities and lower prices. In addition, couples seeking a destination wedding venue may be enticed by the larger banquet facility. This would be a huge blow to Mirasol's revenue, since weddings make up 20% of its bookings from April through October.

Appointment bookings at the spa have dropped, making it difficult for Sand to justify keeping his three aestheticians on staff. In previous years, the spa was a profit center for the resort. Sand feels that the spa is disorganized and suffers from a lack of consistency in atmosphere: depending who is at the desk, the customer experience could range from muted and distant to boisterously friendly.

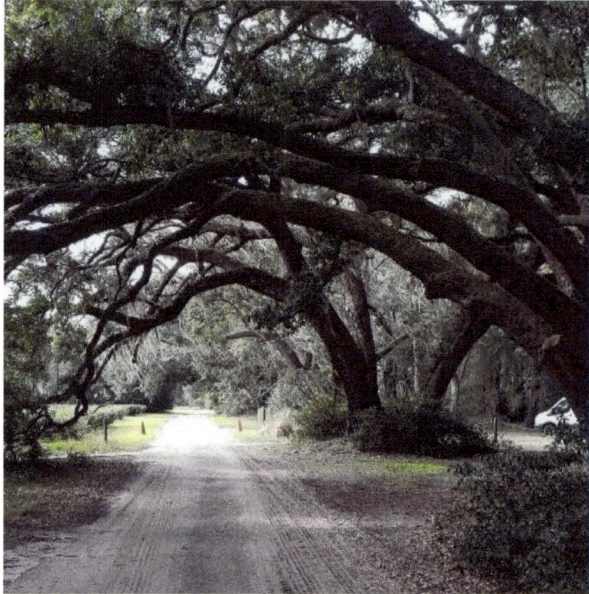

Reina Andersen has been pushing to expand the merchandising business by marketing a line of Mirasol stemware and bar accessories to sell through department stores such as Dilliard's and Nordstrom. She feels this will create new interest for the target market and will be a natural addition to the wedding gift registries of couples married at Mirasol. Both Sand and Greg Shultz, the Controller, see potential in the idea, but are reluctant to make the capital investment for the new line given the current economic climate.

OPERATIONAL ISSUES

According to Eli Guzmán, the Housekeeping Manager, water usage in the on-premises laundry facility has increased by 25% compared to last year. Over the past few months, he has worked with the service contractor to adjust the wash cycle parameters, but the changes have not reduced water usage to target levels. Guzmán has suggested switching to a contractor that leases wash water recycling systems, but Greg Schultz has rejected the idea based on the high installation fee and monthly leasing costs.

The golf cart maintenance garage has experienced parts shortages in the past six months due supplier problems in Japan. This past summer, three of the ten large, 8-passenger golf carts were out of commission, which caused delays in shuttling guests to and from the ferry dock. Chief Engineer Jim Sutton would like to stockpile replacement parts to avoid shortages in the future, but the garage space is already too small for the number of vehicles they service, and parts and tools are often difficult to find.

Inclement weather or equipment problems can cause delays in the ferry schedule, which can leave Mirasol understaffed at crucial times. Sand has adopted an "all hands on deck" approach when these incidents occur, but knows that the quality and speed of the work performed suffers. Last Friday, he and his uncle cleaned 4 rooms when the ferry was two hours late due to engine trouble. Later, his housekeeping staff told him that he had missed vacuuming under the beds, failed to replenish the Tazo tea bags, and did not fold the duvet according to standard.

There has been high absenteeism and significant turnover in the maintenance and housekeeping departments in the past 6 months. Sand believes that the new resort may be offering higher pay. Training new staff members has taken time away from other duties of the Chief Engineer and Housekeeping Manager.

After the sale of Mirasol, the Executive Chef and Banquet Manger decided to remain with Allendale Holdings and were reassigned to a resort on the Gulf Coast of Florida. Since then, Frank Soule, the sous chef, has taken over the Executive Chef and Banquet Manager duties while an outside firm conducts a talent search. Frank has handled his new responsibilities reasonably well, but Sand fears that he may quit due to exhaustion if new staff members are not found quickly.

The quality of service has suffered recently. In the past, an average of 4 complaint cards per month was received by the front desk. In June, 25 complaints were filed by guests. Customers have complained about lengthy waits for the shuttle, special requests that were ignored, incorrect room service orders, inoperable air conditioners and overcharges upon check out. The overcharges have particularly grated on Sand, who wants to make sure this mistake never happens again.

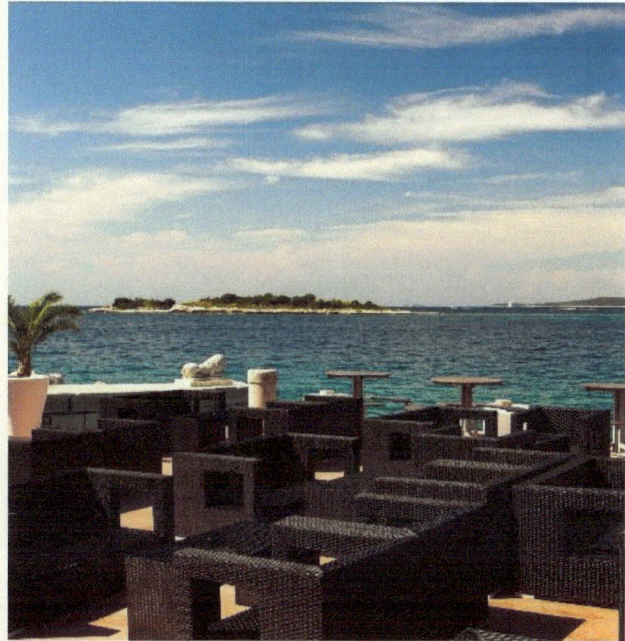

SIX SIGMA IMPLEMENTATION

Sand was involved in quality initiatives at the automotive plant where he previously worked, and had met with the plant's quality manager to discuss data collection and SPC charts a few times. He had tried to incorporate some quality principles into the business operations while he was general manager of the resort back in the Allendale days, but he had neither the experience nor the corporate support to implement a full-fledged program. While attending a hotel and hospitality convention last fall, Sand heard a presentation from an owner of a mid-sized hotel that had decreased costs and increased occupancy by implementing Six Sigma and Lean techniques. Sand began researching Six Sigma and decided to use the DMAIC process to address some of his resort's problems. He himself became Green Belt certified, along with the housekeeping manager and chief engineer.

The other directors and managers have gone through a 2-day Yellow Belt course, led by an outside consultant who is a Six Sigma Black Belt. With the Black Belt's guidance, SIR plans to identify and complete 3-4 Six Sigma projects this year.

Sand Island Resorts - Mirasol Hotel Organizational Chart

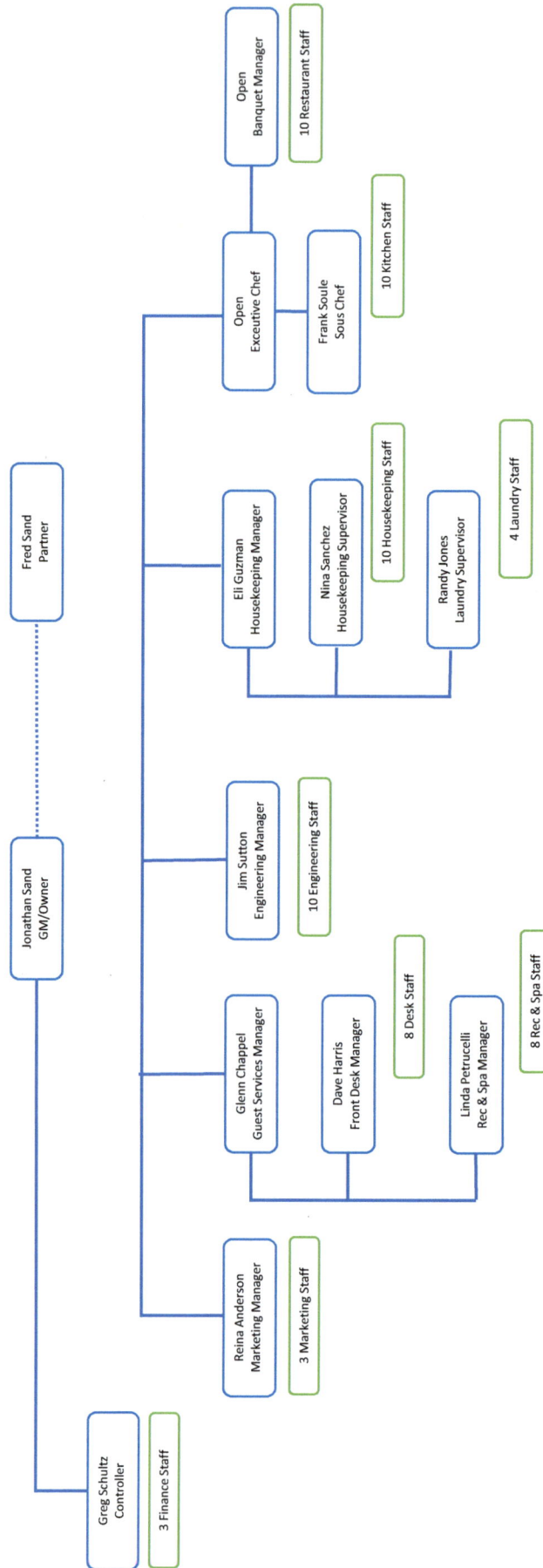

```
                          Jonathan Sand          Fred Sand
                          GM/Owner ............. Partner

Greg Schultz    Reina Anderson    Glenn Chappel      Jim Sutton       Eli Guzman        Open
Controller      Marketing         Guest Services     Engineering      Housekeeping      Executive Chef
                Manager           Manager            Manager          Manager

3 Finance Staff 3 Marketing       Dave Harris        10 Engineering   Nina Sanchez      Frank Soule      Open
                Staff             Front Desk         Staff            Housekeeping      Sous Chef        Banquet Manager
                                  Manager                             Supervisor

                                  8 Desk Staff                        10 Housekeeping  10 Kitchen Staff  10 Restaurant Staff
                                                                      Staff

                                  Linda Petrucelli                    Randy Jones
                                  Rec & Spa Manager                   Laundry Supervisor

                                  8 Rec & Spa Staff                   4 Laundry Staff
```

Data Tables prepared by Greg Schultz, Reina Anderson and Dave Harris

All data from previous year

Occupancy rate by month

January	10%
February	15%
March	20%
April	45%
May	50%
June	75%
July	70%
August	65%
September	65%
October	45%
November	20%
December	30%

Average Length of Stay in days by month

January	1.6
February	1.7
March	2.8
April	3.6
May	2.5
June	3.5
July	3.9
August	3.6
September	3.7
October	1.7
November	1.5
December	2.6

Revenue per Room by month

January	$ 1,395
February	$ 1,890
March	$ 3,255
April	$ 7,628
May	$ 9,300
June	$ 17,438
July	$ 17,360
August	$ 16,120
September	$ 12,188
October	$ 7,324
November	$ 2,400
December	$ 4,883

Percent Bookings by reservation channel

Channel	Percent
Hotel website	54%
Direct Calls	16%
Online booking agencies	30%

Total Room Changeovers by Month

Month	Count
January	78
February	99
March	89
April	150
May	248
June	257
July	223
August	224
September	211
October	328
November	160
December	143

Total complaints by month

Month	Count
January	2
February	2
March	4
April	4
May	5
June	25
July	20
August	24
September	18
October	27
November	15
December	10

Percent complaints per month by room changeover

Month	Percent
January	3%
February	2%
March	4%
April	3%
May	2%
June	10%
July	9%
August	11%
September	9%
October	8%
November	9%
December	7%

Note: A changeover occurs when a guest checks out of a room and a new guest checks in

Lean Six Sigma Introduction

Quiz: Six Sigma is a ...

a) management philosophy
b) systematic approach to problem solving
c) statistical standard of quality
d) all of the above

2

a) Management Philosophy

- Organization-wide deployment and involvement
- Driven from the top down
- Elevated standard of excellence
- Linked to the business' bottom line
- Requires in-depth training of quality tools
- Oriented toward projects

3

b) Systematic Approach to Problem Solving

- Process focused

A process is a sequence of steps that uses inputs and produces a product or service as an output.

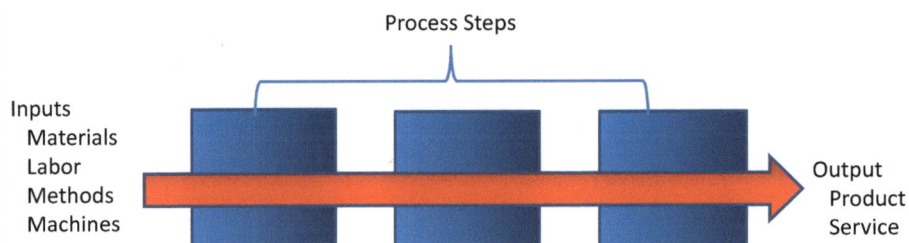

Process Steps

Inputs
Materials
Labor
Methods
Machines

Output
Product
Service

4

Six Sigma Problem Solving Approach

- Problems solved through projects (2 - 9 months)
- Projects follow five clearly defined phases
- Decisions are data-driven

Reduce variation
Decrease defects

Increase customer satisfaction
Improve bottom line results

The Six Sigma Project Method

Customer Requirements

Business Strategy

Company Strategic Goals

Project Selection

c) Statistical Standard of Quality

- A Six Sigma (6 σ) level of quality translates into

   3.4 defects per million opportunities (dpmo)

$$6\sigma$$

Six Sigma: When Great Isn't Good Enough

What does the 6 sigma = 3.4 dpmo standard of quality really mean?

In April 2010, a major hamburger chain served approximately 1.26 billion hamburgers.

- At a 99% quality level = 3.8 sigma

   12,600,000 undercooked hamburgers

- At a 99.99966% quality level = 6 sigma

   4,284 undercooked hamburgers

More Examples

99% Quality = 3.8 sigma	99.99966% Quality = 6 sigma
• 7,011 vehicle breakdowns per **day**	• 870 breakdowns per **year**
• 2,441 boating distress calls per **month**	• 10 distress calls per **year**
• 3,540 dropped babies per **week**	• 5 dropped babies per **month**
• 1,452 surgical mistakes per **day**	• 180 surgical mistakes per **year**

With 1.5 sigma shift

9

A Few Actual Sigma Levels

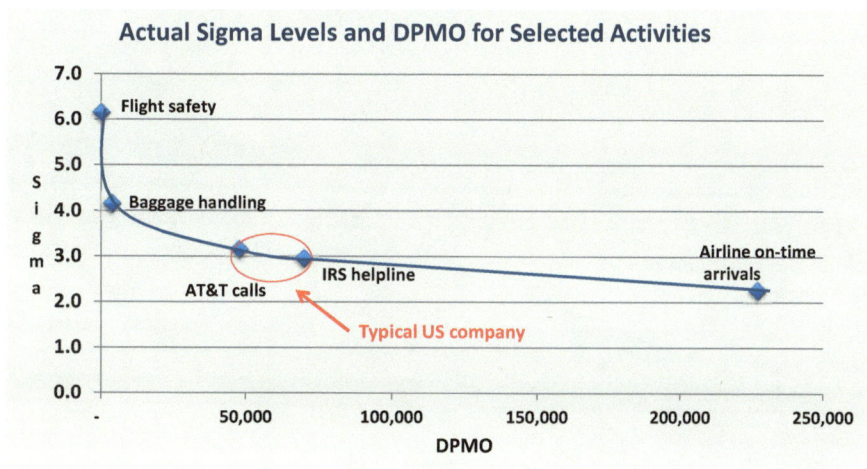

Actual Sigma Levels and DPMO for Selected Activities

With 1.5 sigma shift

10

Quiz: Six Sigma is a ...

a) management philosophy
b) systematic approach to problem solving
c) statistical standard of quality
d) all of the above

Origins of Six Sigma*

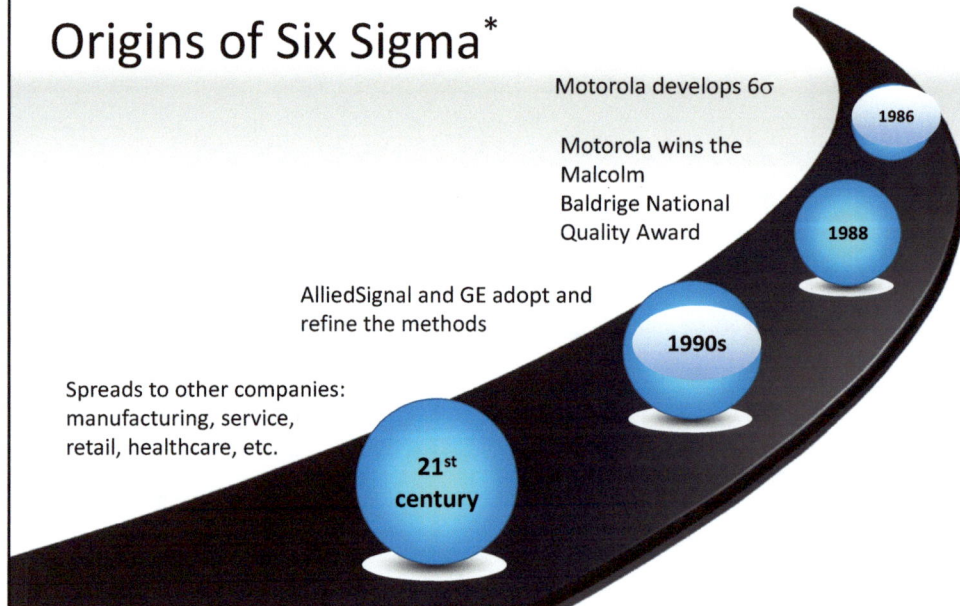

Motorola develops 6σ — **1986**

Motorola wins the Malcolm Baldrige National Quality Award — **1988**

AlliedSignal and GE adopt and refine the methods — **1990s**

Spreads to other companies: manufacturing, service, retail, healthcare, etc. — **21st century**

*Six Sigma is a registered trademark and service mark of Motorola, Inc.

Black Belt
- Works full time leading Six Sigma projects.
- Knowledge of statistical methodology and team dynamics.
- Trains and mentors Green Belts

Green Belt
- Retains regular position, but works part-time on Six Sigma
- May lead small teams.
- Knowledge of many Six Sigma tools.

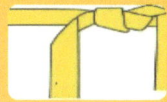

Yellow Belt
- Works part-time as a Six Sigma team member.
- Knowledge of the Six Sigma process and some tools.

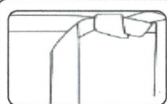

White Belt.
- Has awareness of Six Sigma concepts.
- May or may not work on a Six Sigma team

Why the Karate Theme?

- We can thank Dr. Mikel Harry at Motorola

A manager told him the Six Sigma tools were "kicking the hell out of variation."

14

More Six Sigma Roles

- Master Black Belt
  - Full time position
  - Trains and mentors Black Belts
  - Works with management to choose Six Sigma projects
  - Advanced knowledge of statistical techniques

- Champion or Sponsor
  - Top-level manager familiar with Six Sigma
  - Supports teams by providing resources and removing barriers

- Executive
  - Shows support of Six Sigma through communication and actions
- Process Owner
  - Team member or leader
  - Manager responsible for all aspects of a process with the authority to make changes
  - Should be at least a Green Belt

15

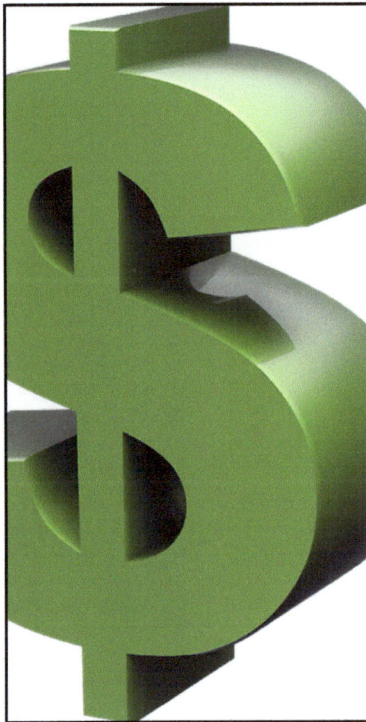

Six Sigma Features

- Enterprise wide
- Top down
- Data driven
- Project oriented
- Extremely high standard for quality

- **BOTTOM LINE RESULTS**

  Something top management can get behind

16

According to the Six Sigma Academy...

Black Belts save companies approximately $230,000 per project and can complete 4 to 6 projects per year.

General Electric, one of the most successful companies implementing Six Sigma estimated benefits of around **$10 billion** during the first five years of implementation.

Some Six Sigma Results

17

Lean

✓ Common sense approach
✓ Hands on training
✓ Little to no math
✓ Quick, dramatic results

• Emphasis on <u>reducing waste</u>, continuous improvement

Transportation
Inventory
Motion
Waiting
Overproduction
Over-processing
Defects
Skills

18

Best of Both Worlds

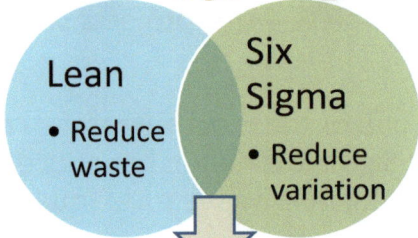

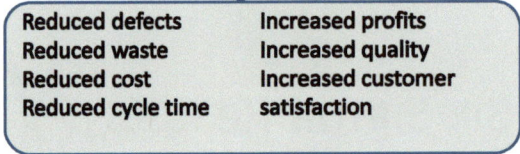

Lean
- Reduce waste

Six Sigma
- Reduce variation

Reduced defects   Increased profits
Reduced waste   Increased quality
Reduced cost   Increased customer
Reduced cycle time   satisfaction

19

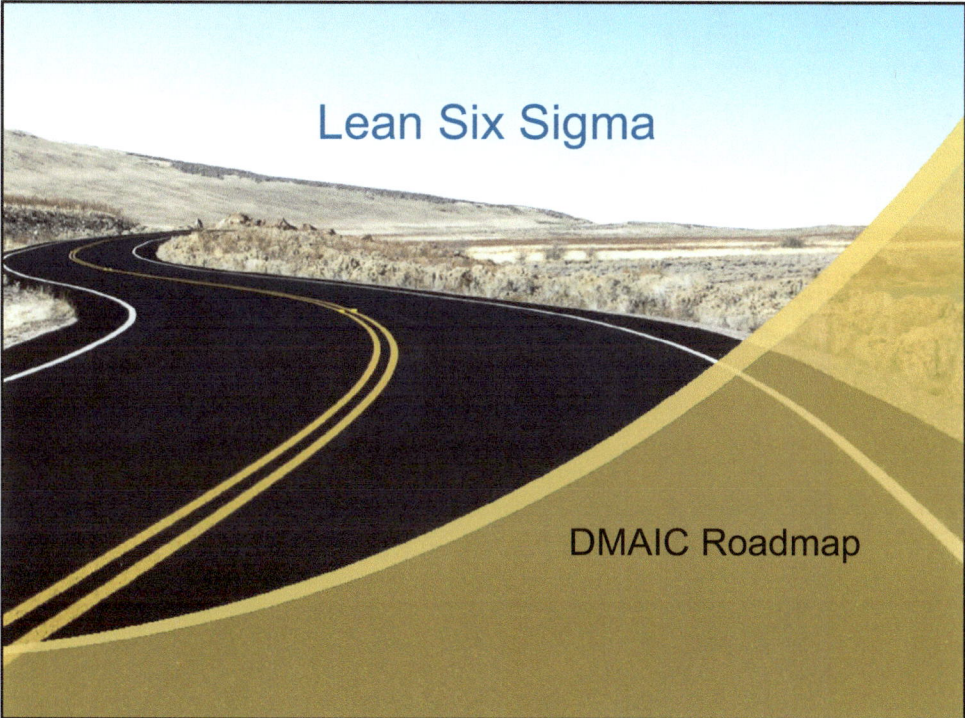

Lean Six Sigma

DMAIC Roadmap

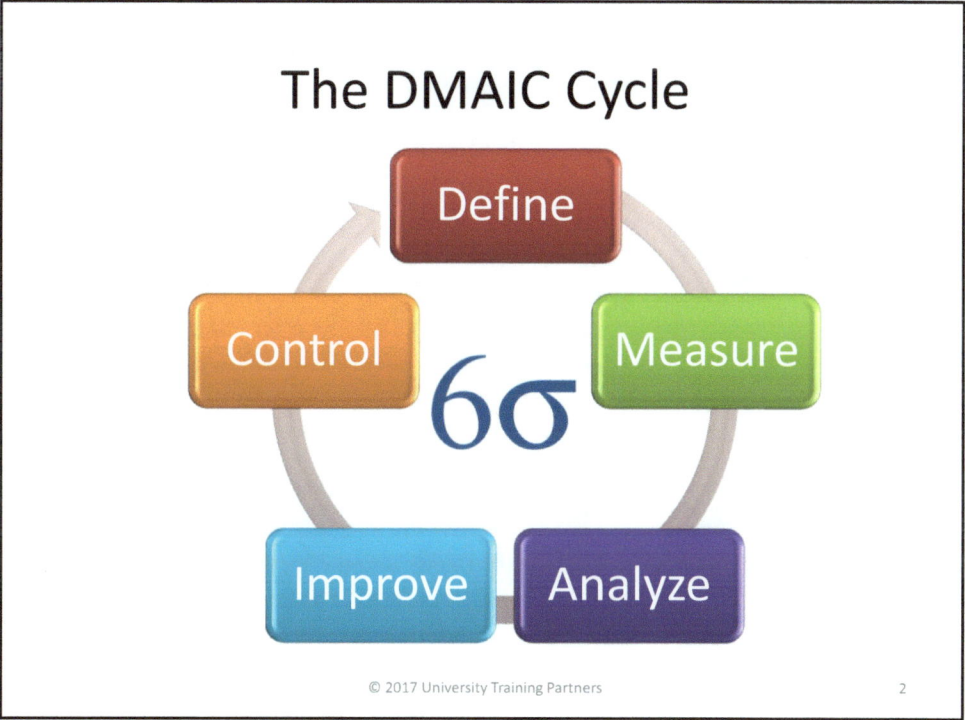

The DMAIC Cycle

Define

Measure

Analyze

Improve

Control

6σ

2

DMAIC Activities by Phase

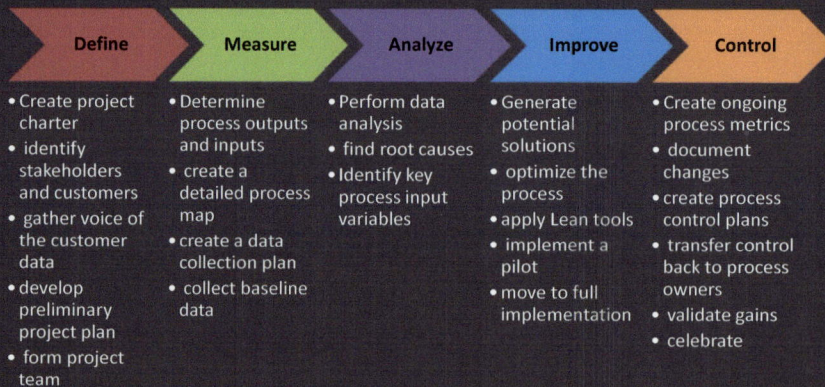

Define	Measure	Analyze	Improve	Control
• Create project charter • identify stakeholders and customers • gather voice of the customer data • develop preliminary project plan • form project team	• Determine process outputs and inputs • create a detailed process map • create a data collection plan • collect baseline data	• Perform data analysis • find root causes • Identify key process input variables	• Generate potential solutions • optimize the process • apply Lean tools • implement a pilot • move to full implementation	• Create ongoing process metrics • document changes • create process control plans • transfer control back to process owners • validate gains • celebrate

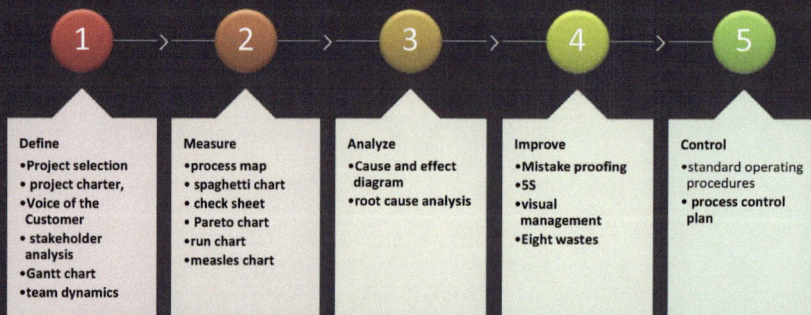

© 2017 University Training Partners

3

DMAIC Tools

1	2	3	4	5
Define • Project selection • project charter, • Voice of the Customer • stakeholder analysis • Gantt chart • team dynamics	**Measure** • process map • spaghetti chart • check sheet • Pareto chart • run chart • measles chart	**Analyze** • Cause and effect diagram • root cause analysis	**Improve** • Mistake proofing • 5S • visual management • Eight wastes	**Control** • standard operating procedures • process control plan

© 2017 University Training Partners

4

Capturing the Voice of the Customer

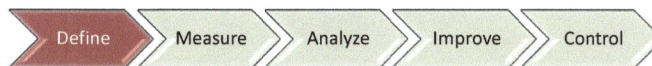

Define → Measure → Analyze → Improve → Control

Who is the Customer?

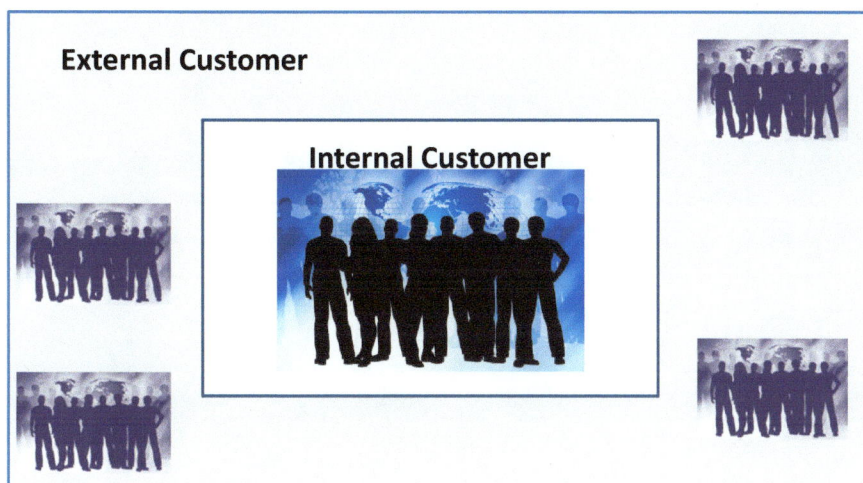

External Customer

Internal Customer

2

VOC

- Listen to the Voice of the Customer in order to

  - Discover what customers care about (and what they are willing to pay for)

  - Align corporate and project goals and priorities with customer needs

3

Capturing the Voice of the Customer

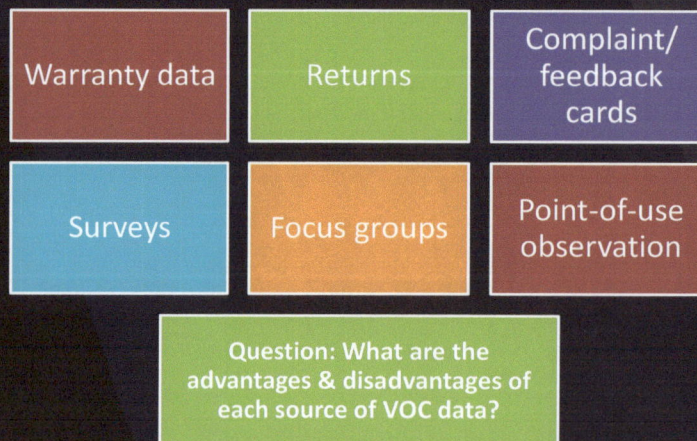

| Warranty data | Returns | Complaint/ feedback cards |
| Surveys | Focus groups | Point-of-use observation |

Question: What are the advantages & disadvantages of each source of VOC data?

© 2017 University Training Partners

4

Translating the VOC into ...

CTQ = Critical-to-Quality Requirements

Dashboards

Kano Analysis

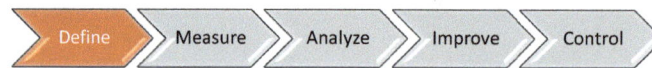

Define > Measure > Analyze > Improve > Control

Kano Analysis

- Customers have specific needs and wants.

These can be spoken or unspoken. They can even be unknown to them.

Spoken

- **Satisfiers** – performance features
- **Must-haves** – basic requirements

Unspoken

- **Delighters** – excitement features

Unknown

The Kano Model

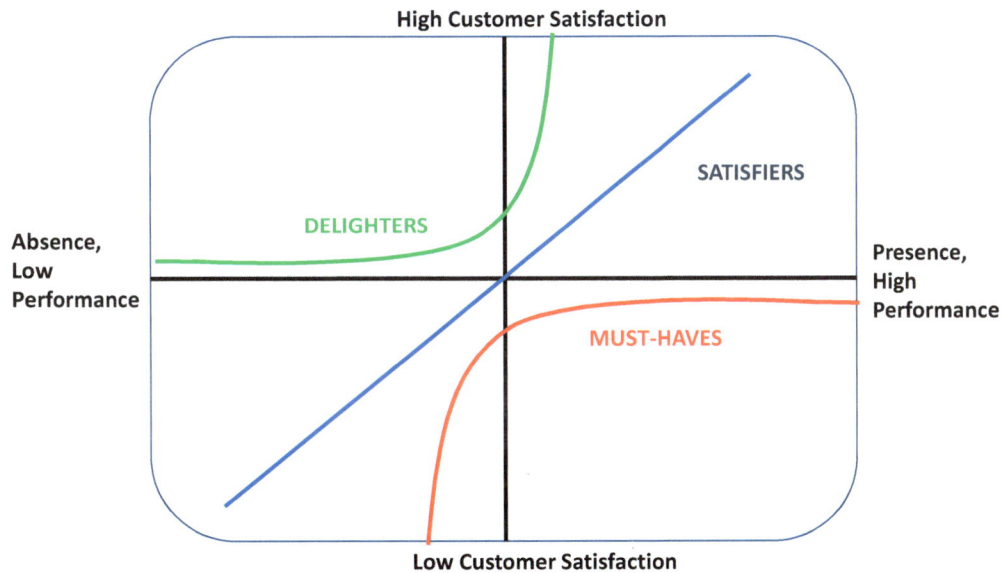

Determining Kano Classifications

Kano categories can be determined through specialized surveys using two questions for each feature.

Presence, or *functional question*:
If the bed linens in a hotel are of high quality, how do you feel?
- o Happy, I **like** it that way
- o Neutral, I **expect** linens to be high quality
- o Neutral, high quality linen is not important to me
- o I can **live with** it
- o Unhappy, I **dislike** it that way

Absence or *dysfunctional question*:
If the bed linens in a hotel are not of high quality, how do you feel?
- o Happy, I don't **like** high quality linens
- o Neutral, I don't **expect** high quality linens
- o Neutral, high quality linen is not important to me
- o I can **live with** it
- o Unhappy, I **dislike** low quality linens

Classifying Kano Features

Questionable, inconsistent

Delighter

One-dimensional (satisfier)

Must have

Indifferent

Reverse

Customer Requirements		Dysfunctional				
		Like	Expect	Neutral	Live with	Dislike
F u n c t i o n a l	Like	Q	D	D	D	O
	Expect	R	I	I	I	M
	Neutral	R	I	I	I	M
	Live with	R	I	I	I	M
	Dislike	R	R	R	R	Q

M – Must have
O – One Dimensional
Delighter

I – Indifferent, customer doesn't really care about the feature
Q – Questionable, customer answers are inconsistent
R – Reverse, features actually make customer feel worse

5

Using Kano Survey Results

Classifying Kano Features

Customer Requirement		Dysfunctional				
		Like	Expect	Neutral	Live with	Dislike
F u n c t i o n a l	Like	Q	D	D	D	O
	Expect	R	I	I	I	M
	Neutral	R	I	I	I	M
	Live with	R	I	I	I	M
	Dislike	R	R	R	R	Q

Questionable, inconsistent

Delighter

One-dimensional (satisfier)

Must have

Indifferent

Reverse

M – Must be
O – One Dimensional
D – Delighter

I – Indifferent, customer doesn't really care about the fea...
Q – Questionable, customer answers are inconsistent
R – Reverse, features actually make customer feel worse

2

Individual Responses

Calculated cells	
INDIVIDUAL RESULTS - Your name here	
Feature	**Kano Designation**
High quality bed linens	S
High quality restaurant meals	S
Swim-up bar	I
Room service	I
Spa services such as mani/pedis	D
Nightly entertainment	I
Water sports	D
A high cleanliness score	M
High quality toiletries	I
Free wifi	M

Kano Survey Results

GROUP TALLY

Feature	Must Have	Satisfier	Delighter	Indifferent	Reverse	Questionable
High quality bed linens	35%	33%	12%	20%	0%	0%
High quality restaurant meals	21%	22%	25%	32%	0%	0%
Swim-up bar	0%	0%	15%	75%	10%	0%
Room service	10%	7%	0%	65%	8%	0%
Spa services such as mani/pedis	0%	11%	15%	70%	4%	0%
Nightly entertainment	11%	6%	16%	67%	0%	0%
Water sports	0%	6%	34%	60%	0%	0%
A high cleanliness score	53%	45%	0%	0%	0%	2%
High quality toiletries	19%	11%	55%	15%	0%	0%
Free wifi	67%	19%	7%	7%	0%	0%

4

Project Selection and Scope

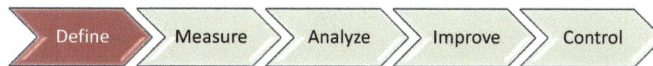

Define ▷ Measure ▷ Analyze ▷ Improve ▷ Control

The Six Sigma Project Method

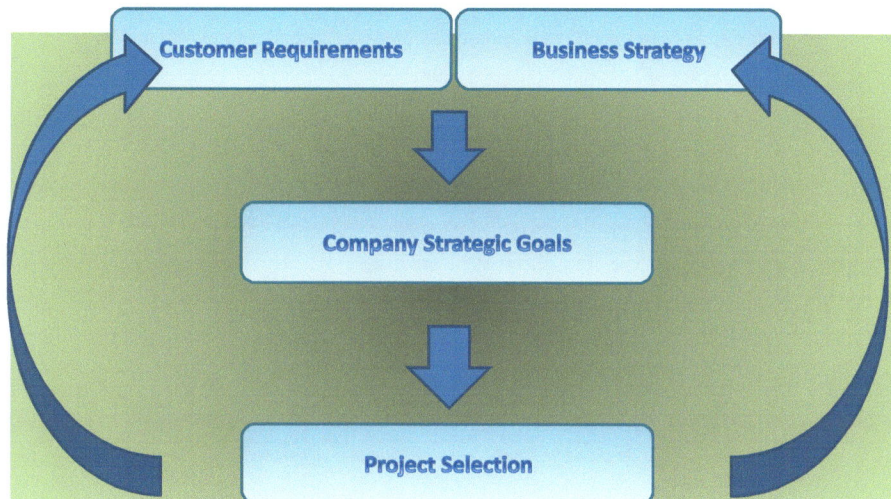

Customer Requirements | **Business Strategy**

Company Strategic Goals

Project Selection

Selecting Projects

- Performed by a steering committee

- Considerations
  - Does the project tie into the company strategic plan?
  - Does the project address customer needs?
  - What is the probable duration of the project?
  - What is the expected cost-benefit ratio?
  - What is the project's level of complexity?
  - Does the company have the resources available for the project?

Selecting Projects

Voice of the Business | Voice of the Customer | Voice of the Process

Selecting Projects

| Environmental Health and Safety | Government, Regulatory | Voice of the Employees |

Prioritization Matrix Example

Choosing Among Office Supply Vendors

Use a **prioritization matrix** to compare various options based on set criteria

Criteria	Selection	Cost	Speed	Support
Relative Importance	25%	40%	20%	15%
Option				
Vendor A	1	1.5	3	1
Vendor B	3	3	2	2
Vendor C	2	1.5	1	3

Criteria	Selection	Cost	Speed	Support	
Relative Importance	25%	40%	20%	15%	
Option					Total
Vendor A	0.25	0.6	0.6	0.15	**1.60**
Vendor B	0.75	1.2	0.4	0.3	**2.65**
Vendor C	0.5	0.6	0.2	0.45	**1.75**

Project Scope

- Project should take 2-9 months to complete
- Don't try to solve world hunger…
  - "Improve customer satisfaction"
  - "Reduce warranty claims"
- Don't make scope so narrow that it prevents the root cause from being found
  - "Decrease wait times for customers by increasing the number of support staff."

The Project Charter

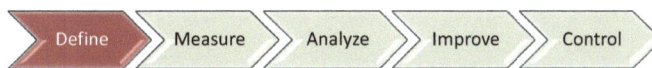

Define → Measure → Analyze → Improve → Control

Project Charter

- Outlines project purpose, benefits, scope, expected results
- Components
  - Project Title
  - Problem Statement
  - Goal Statement
  - Baseline Metrics and Goal Metrics
  - Team members and roles
  - Milestones and completion dates
  - Resources needed

2

Crafting the Problem Statement

- Over the past six months, the number of surface defects on door panels has increased from 5/1000 to 60/1000, resulting in increased rework and warranty claims.

- How long has the problem existed?
- What measurable item is affected?
- What is the performance gap?
- What is the business impact?

3

Problem Statement Cautions

- Causes of the problem or solutions do not belong in the problem statement
  - That's what the project is for
  - If solution is obvious, then is it really a project?
- Make sure item is measurable
- Need data in the problem statement, not conjecture

4

Problem Statement Examples

- Customer complaints have always been higher than they should be, resulting in lost sales.

- Reduce sealing defects from 5% to 1% by replacing the sealing machine with an improved model.

- Over the past year, hotel bookings have decreased by 10% due to rude employees, resulting in a $400,000 budget shortfall.

Dr. Deming stated that 94% of all problems, defective goods or services came from the system, not from a careless worker or a defective machine.

5

What about these?

- Over the past two years, customer dining complaints have increased by 20%, resulting in fewer repeat customers.

- Order fulfillment time for product X has increased from 3 weeks to 12 weeks in the past year, allowing our competitor to increase its market share.

How long
Measurable item
Performance gap
Business impact

6

How Did You Do?

- Over the past two years, customer dining complaints have increased by 20%, resulting in fewer repeat customers.

- Order fulfillment time for product X has increased from 3 weeks to 12 weeks in the past year, allowing our competitor to increase its market share.

How long
Measurable item
Performance gap
Business impact

7

Goal Statement

- SMART Goal
  - **S** – **S**pecific
  - **M** –**M**easurable
  - **A** – **A**chievable
  - **R** – **R**ealistic
  - **T** – **T**ime bound

- "Reduce surface defects on doors from 60/1000 to 5/1000 within 6 months."

8

Building the Gantt Chart

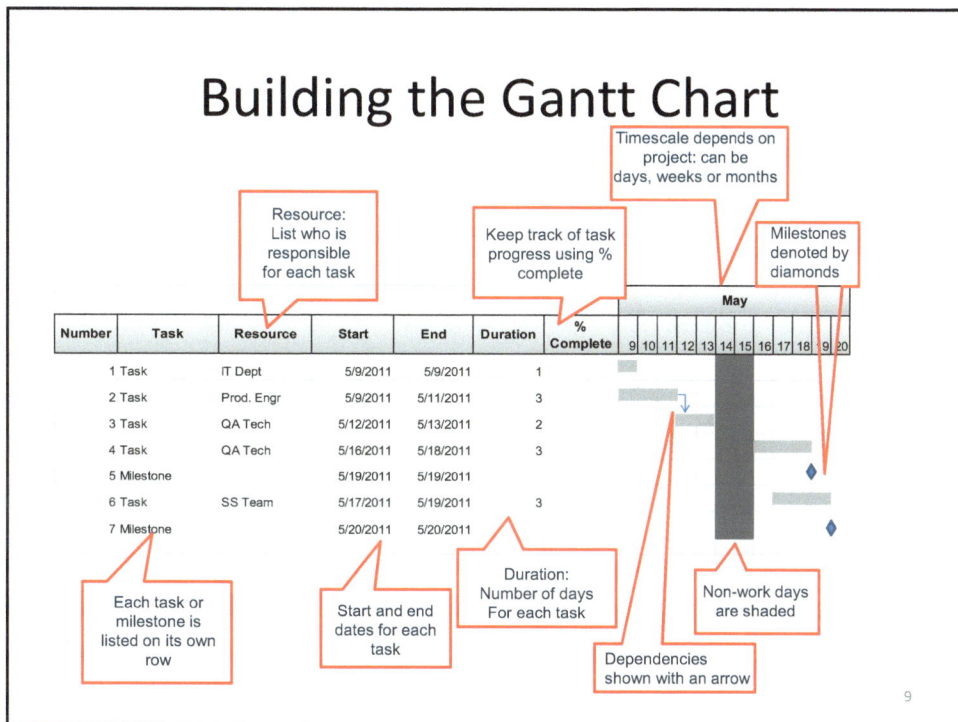

Timescale depends on project: can be days, weeks or months

Resource: List who is responsible for each task

Keep track of task progress using % complete

Milestones denoted by diamonds

Number	Task	Resource	Start	End	Duration	% Complete	May
							9 10 11 12 13 14 15 16 17 18 9 20
1	Task	IT Dept	5/9/2011	5/9/2011	1		
2	Task	Prod. Engr	5/9/2011	5/11/2011	3		
3	Task	QA Tech	5/12/2011	5/13/2011	2		
4	Task	QA Tech	5/16/2011	5/18/2011	3		
5	Milestone		5/19/2011	5/19/2011			
6	Task	SS Team	5/17/2011	5/19/2011	3		
7	Milestone		5/20/2011	5/20/2011			

Each task or milestone is listed on its own row

Start and end dates for each task

Duration: Number of days For each task

Dependencies shown with an arrow

Non-work days are shaded

9

Project Milestone & Tollgates

- Milestones include each phase of the DMAIC process

- Tollgate reviews are held after each phase is completed and before a new phase can begin

10

Stakeholder Analysis

Why do projects fail?

- Too many people on the team
- Wrong people chosen for team
- Lack of communication/ poor communication
- Failure to prepare for change

To paraphrase Tolstoy:

All successful projects are alike; each failed project fails in its own way.

Why do projects fail?

- Too many **people** on team
- Wrong **people** chosen for team
- Lack of communication/ poor communication
- Failure to prepare for change

> To paraphrase Tolstoy:
>
> Successful project are all alike; every failed project fails in its own way.

Stakeholder analysis steps

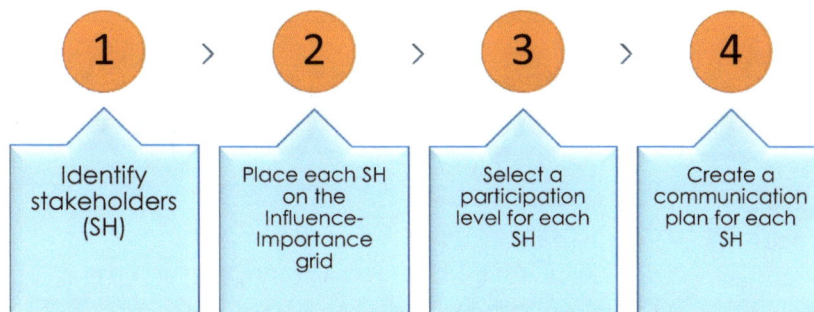

1 > **2** > **3** > **4**

1	2	3	4
Identify stakeholders (SH)	Place each SH on the Influence-Importance grid	Select a participation level for each SH	Create a communication plan for each SH

Stakeholder defined

A stakeholder is anyone who has a vested interest in a project or who can effect or be affected by an action or change.

- Who might receive benefits or experience negative effects?
- Who might be forced to make changes or change behavior?
- Who has goals that align/conflict with the project goal?
- Who has responsibility for action or decisions?
- Who has resources or knowledge that are important to the project?
- Who has expectations for this project?

Influence – Importance grid

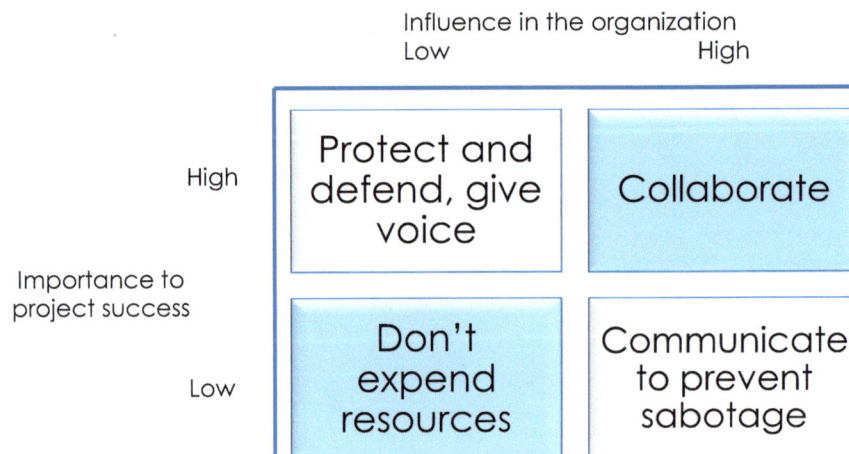

Influence in the organization
Low   High

	Low	High
High (Importance to project success)	Protect and defend, give voice	Collaborate
Low	Don't expend resources	Communicate to prevent sabotage

Influence – Importance grid

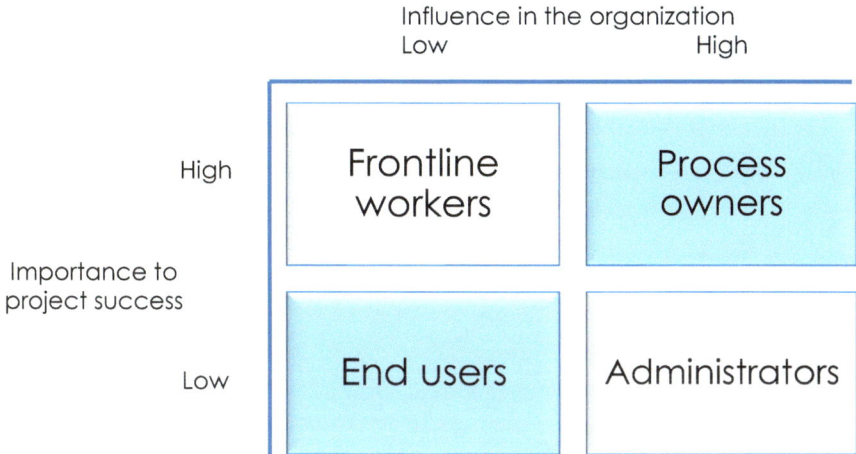

Influence in the organization

	Low	High
High	Frontline workers	Process owners
Low	End users	Administrators

Importance to project success

Stakeholder participation levels

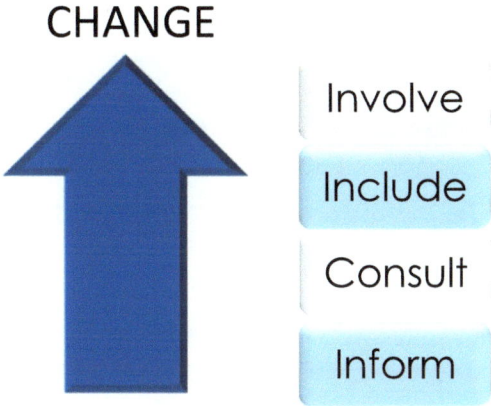

CHANGE

- Involve
- Include
- Consult
- Inform

Stakeholder participation levels

CHANGE

Involve	Have SH in on the team
Include	Invite SH to meetings and include in decision making
Consult	Ask SH to weigh in on certain matters & advise team
Inform	Let SH know what took place

Stakeholder communication plan

| Speak to informally as needed | Send copy of meeting minutes | Meet with regularly |

| Invite to team meetings | Other |

Who will assure communication is performed?

Outcomes

- Listing of Stakeholders
- Behavior approach based on Influence-Importance chart
- Participation level selected based on amount of change required (Involve = Team)
- Communication plan developed for each stakeholder
- Assignment for delivering communication

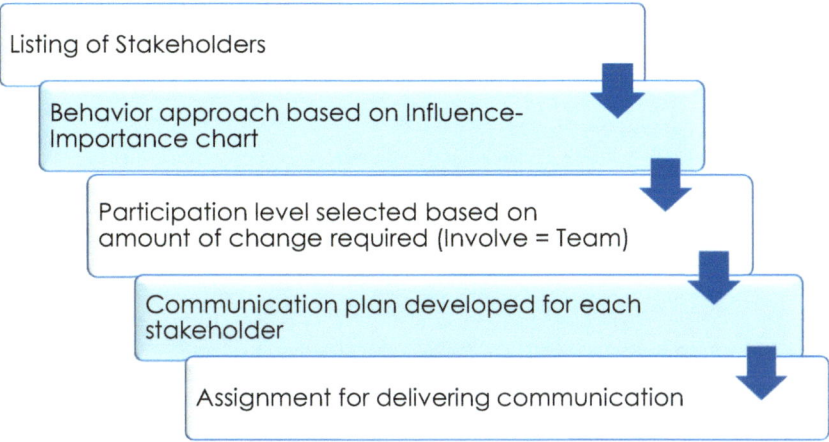

Stakeholder analysis template

Stakeholder Name or Job Title	Why a Stakeholder?						
	Will receive benefits/ negative effects	Will be forced to make changes	Has goals that align/ conflict with project	Has responsibility for action/ decisions	Controls important resources	Has useful expertise	Has expectations for project

Change required

Influence/Importance			Participation Level				Communication Plan					
Influence in the organization Enter L or H only	Importance to the project success L/H	Behavior Approach	Inform	Consult	Include	Involve	Speak with informally as needed	Send copy of meeting minutes	Invite to team meetings	Meet with regularly	Other (describe)	Who is responsible for communication?
L	H	Protect & defend										

Six Sigma Teams

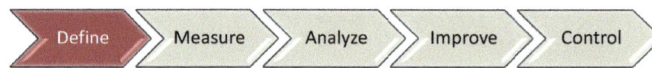

Define ▶ Measure ▶ Analyze ▶ Improve ▶ Control

Six Sigma Project Teams

Should be made up of 5-7 members who are knowledgeable about the process

Other members may serve on the team for a short while as their skill sets are needed

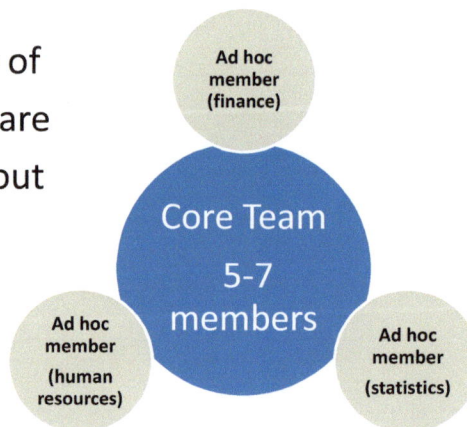

Ad hoc member (finance)

Core Team 5-7 members

Ad hoc member (human resources)

Ad hoc member (statistics)

2

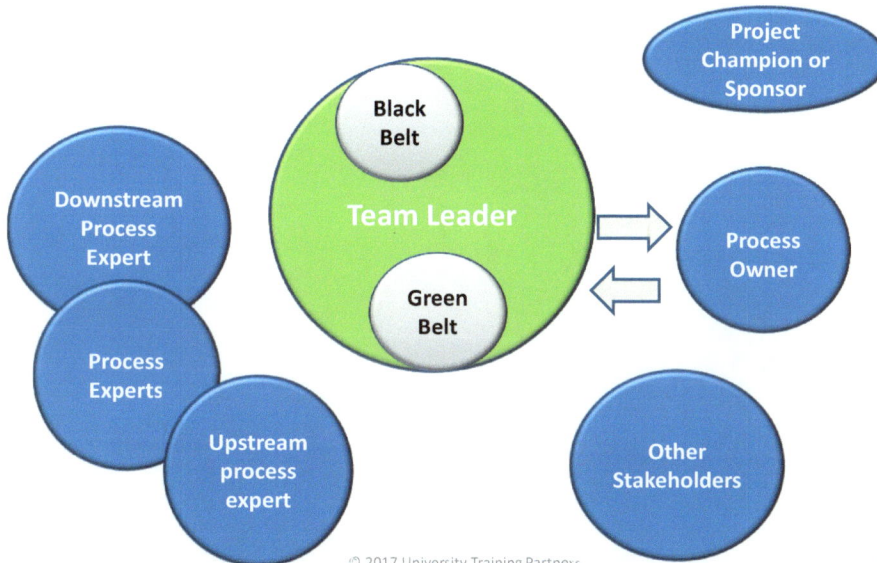

Core Team Members

Black Belt

Team Leader

Green Belt

Project Champion or Sponsor

Downstream Process Expert

Process Experts

Upstream process expert

Process Owner

Other Stakeholders

© 2017 University Training Partners

3

Launching Teams

Align the purpose and goals of the team directly to the project charter

Provide basic team-building training

Publish a schedule for team meetings & have members commit to schedule

Review deliverable assignments and due date at the close of each meeting; publish minutes

Team success depends on management support & the support of the sponsor

© 2017 University Training Partners

4

Team Charter

- The team charter helps the group agree upon ground rules together
- Team expectations are clearly defined
- All members sign off on rules

Team Growth Stages

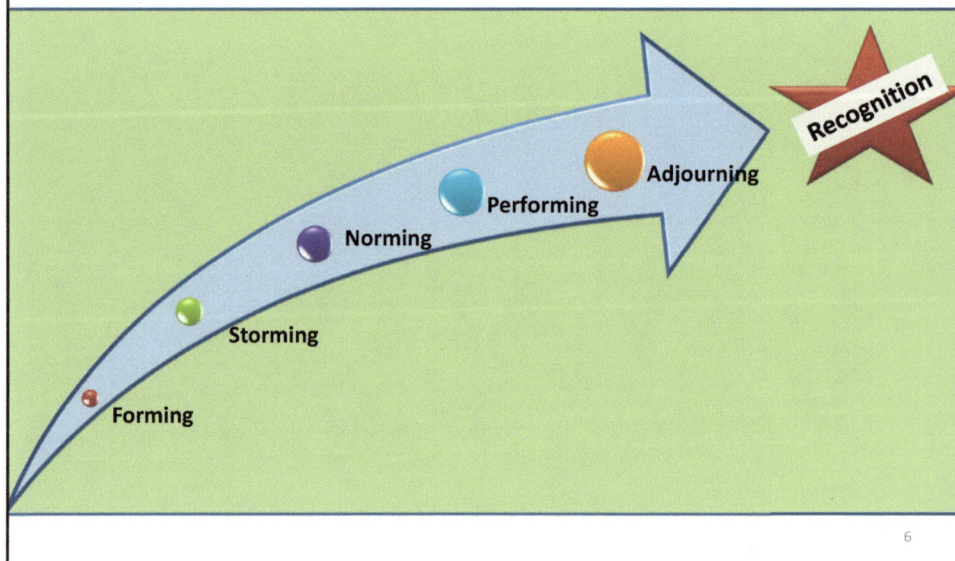

Forming

Storming

Norming

Performing

Adjourning

Recognition

6

When the team is …

- **Forming**, members struggle to understand the goal and how they will contribute
- **Storming**, members express their own opinions and ideas, often disagreeing
- **Norming**, members begin to understand the need to operate as a <u>team</u>
- **Performing**, team members work together to achieve a common goal
- **Adjourning**, a final meeting is held to discuss & summarize project decisions and wrap up loose ends

7

Cost of Quality

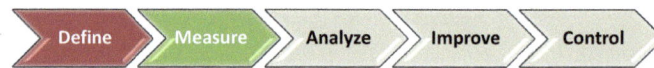

Define | Measure | Analyze | Improve | Control

1

Quality Costs

Internal Failure	External Failure
Appraisal	**Prevention**

2

Quality Costs

Internal Failure

- Scrap
- Rework
- Retest
- Downtime
- Yield losses
- Disposition

Source: Juran and Gryna, Quality Planning and Analysis, 2nd edition, McGraw Hill, 1980, pp. 14-16

3

Quality Costs

External Failure

- Complaint adjustment
- Returned material
- Warranty charges
- Loss of good will

Source: Juran and Gryna, Quality Planning and Analysis, 2nd edition, McGraw Hill, 1980, pp. 14-16

4

Quality Costs

Appraisal

- Incoming material inspection
- Inspection and test
- Maintaining test equipment
- Materials consumed

Source: Juran and Gryna, Quality Planning and Analysis, 2nd edition, McGraw Hill, 1980, pp. 14-16

5

Quality Costs

- Training
- Process control
- Mistake proofing
- Visual management

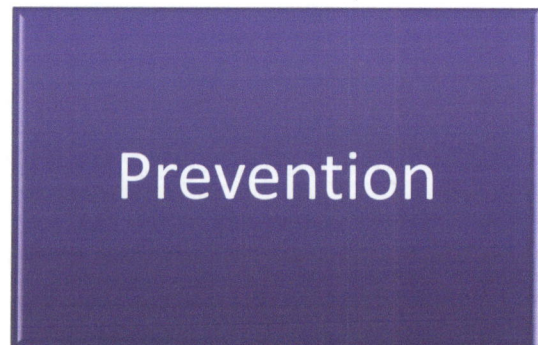

Prevention

Source: Juran and Gryna, Quality Planning and Analysis, 2nd edition, McGraw Hill, 1980, pp. 14-16

6

At First, as Appraisal & Prevention Costs Increase…

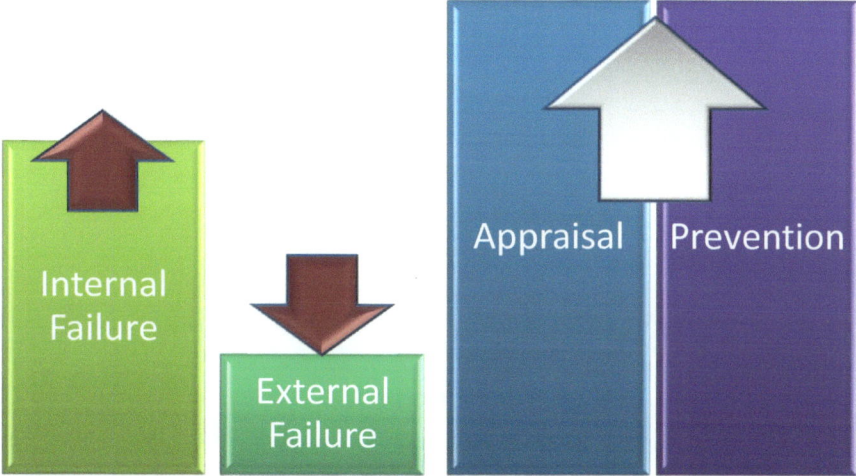

Then, Internal Failures Decrease…

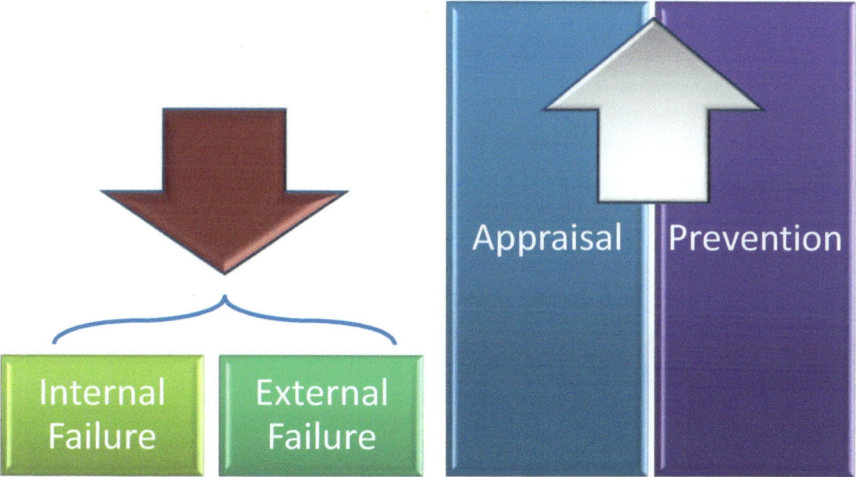

Finally, Preventive Methods Reduce the Need for Appraisal

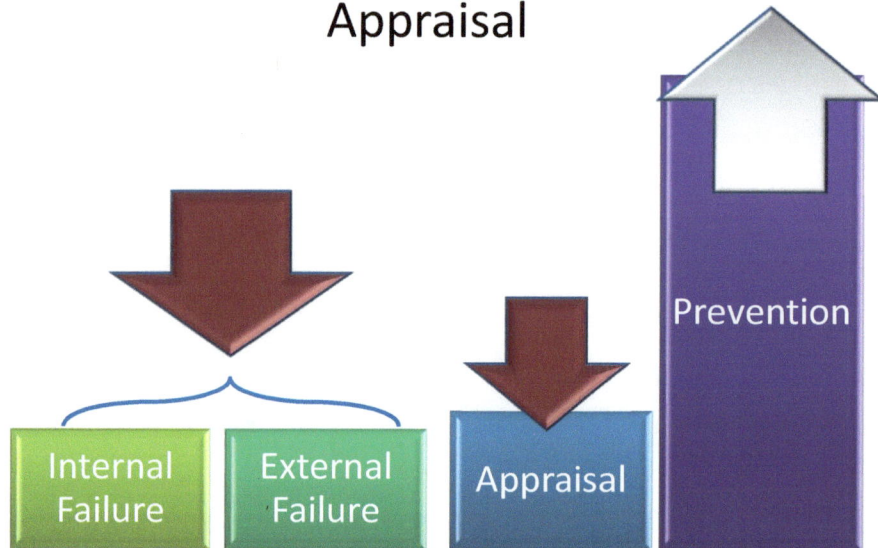

| Internal Failure | External Failure | Appraisal | Prevention |

9

Introduction to Process Maps

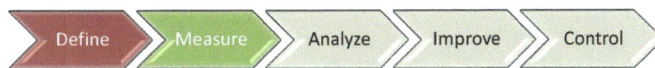

Define → Measure → Analyze → Improve → Control

Process Defined

Recall:

A process is a step or a sequence of steps that uses inputs and produces a product or service as an output.

2

Creating a Process Map

1. Walk the process!   **Go to Gemba**
2. Note process boundaries as defined by project
3. Team writes process steps on post-it notes
4. Team arranges steps in sequence
5. Make any necessary adjustments
6. Number tasks sequentially
7. Transfer map to paper or computer

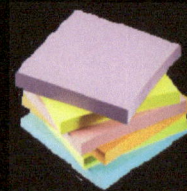

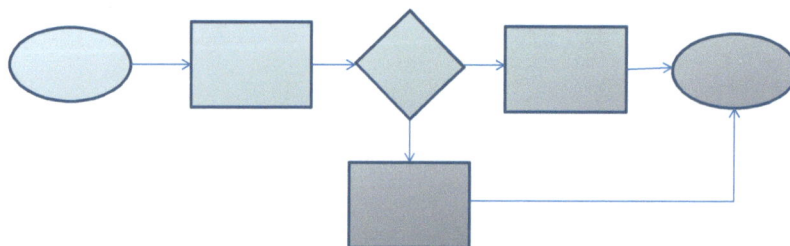

© 2017 University Training Partners

3

Three Basic Map Symbols

Start and End Points   Activity   Decision

© 2017 University Training Partners

4

Mapping Tips

- Make sure the level of detail matches your purposes
- Use a cross-functional team to create the map
- Concentrate on documenting the process, not using the right symbols
- Capture ancillary ideas brought up in the map meeting and place in a Parking Lot

Swim Lane Chart

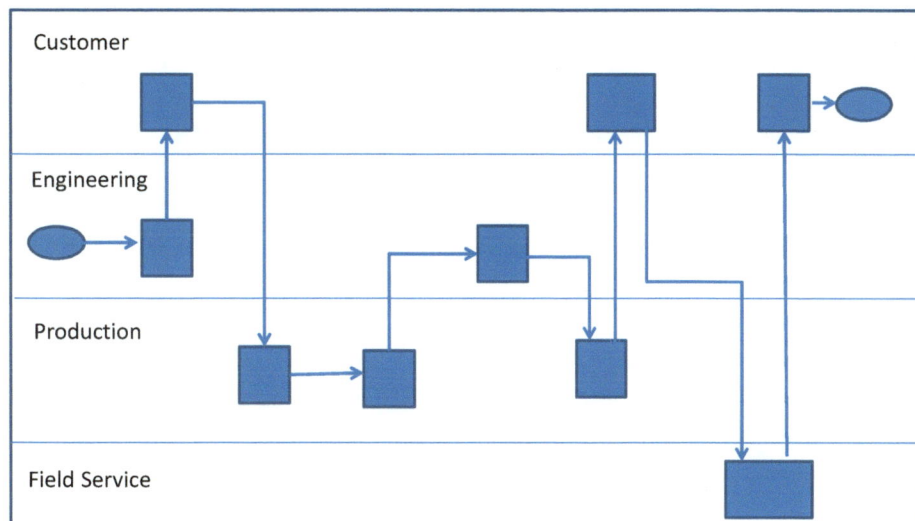

Customer

Engineering

Production

Field Service

6

Spaghetti Diagram

1. Indicate the location of each process step on a diagram of the workspace
2. Connect sequential steps with a line (the spaghetti)

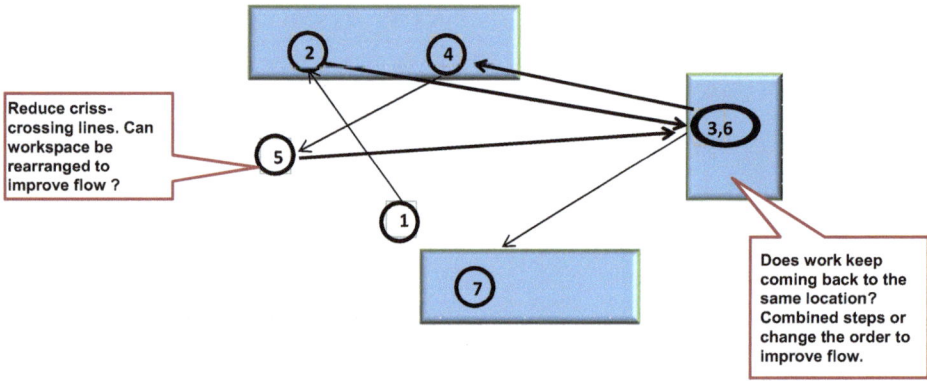

Reduce criss-crossing lines. Can workspace be rearranged to improve flow ?

Does work keep coming back to the same location? Combined steps or change the order to improve flow.

7

Graphical Analysis Tools

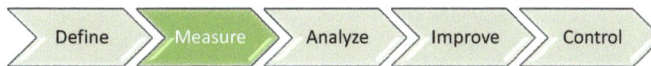

Define > Measure > Analyze > Improve > Control

Check Sheets

- A **Check Sheet** is a data collection tool for counting occurrences of specific events

Identify and write down each event type on its own line. Examples: Types of defects, types of customer complaints

Event Type	Counts
Event Type A	‖‖‖ ‖‖‖
Event Type B	‖‖‖ ‖‖‖ IIII
Event Type C	II
Event Type D	‖‖‖
Event Type E	I
Event Type F	IIII
Event Type G	I

Leave room to write in other event types that occur

2

Pareto Charts

- Named for Vilfredo Pareto
  - Used the 80/20 rule to describe
    wealth distribution in Europe
- Joe Juran applied the
  "Pareto Principle" to quality issues
  - *The Vital few and the trivial many*
  - 80% of the problem can be attributed to 20% of the causes
  - Helps us prioritize our focus to get maximum benefit

3

SIR Case Example
Check Sheet & Pareto Chart

Customer complaint types, August Total = 24

Dept	Complaint	Tally
Engineering	AC/heat	ⅢⅢ I
Desk	Room not ready at check-in	ⅢⅢ III
Desk	Overcharge at check-out	
Housekeeping	Special requests not honored	III
Guest services	Shuttle	IIII
Guest services	Rude employee	I
Restaurant	Room service order incorrect/late	II
Spa	Poor service	

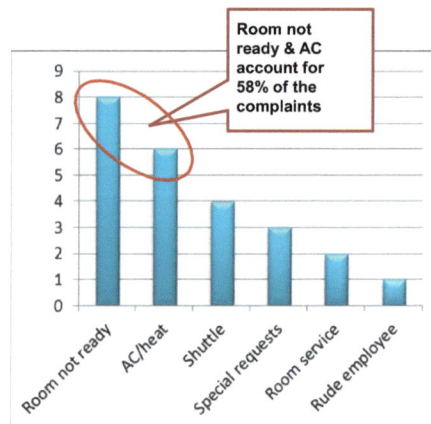

Let's replicate this chart in Excel

> Room not ready & AC account for 58% of the complaints

Run Charts

- A **Run Chart** tracks how a measure or variable performs over time.
- A Run Chart has one variable which is always plotted against *TIME*

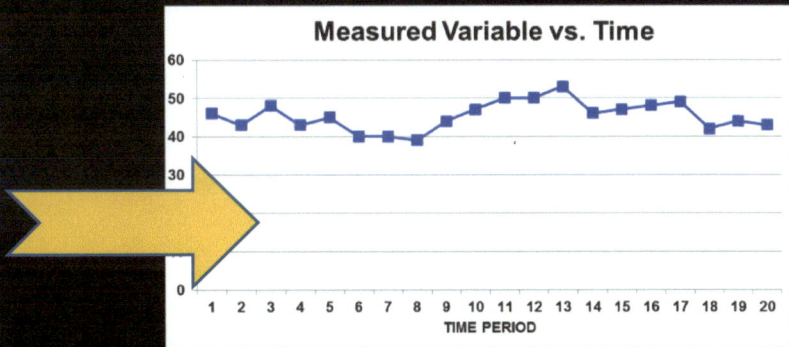

Measured Variable vs. Time

© 2017 University Training Partners

5

Run Chart Patterns and Trends

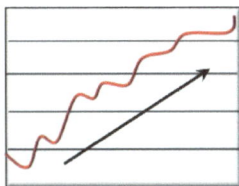

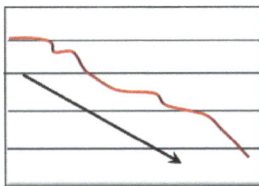

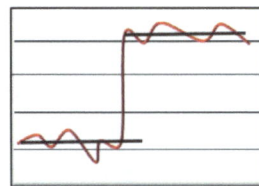

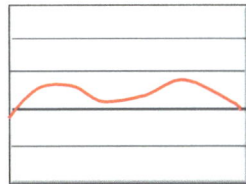

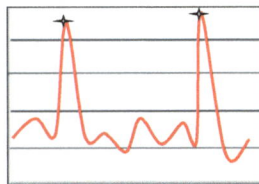

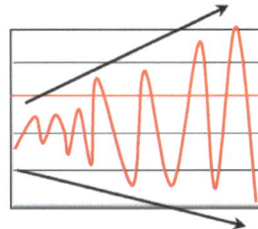

6

SIR Case Example
Guest Complaints By Month

Run Chart Analysis
Data was available for 2011

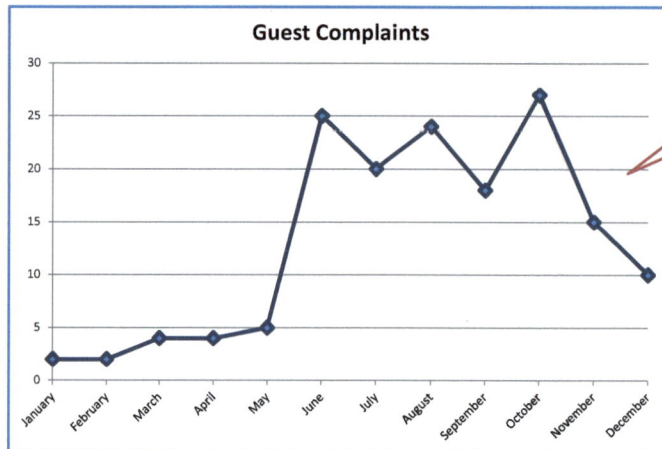

What patterns can you detect?

7

SIR Case Example
Occupancy by Month

Run Chart Analysis

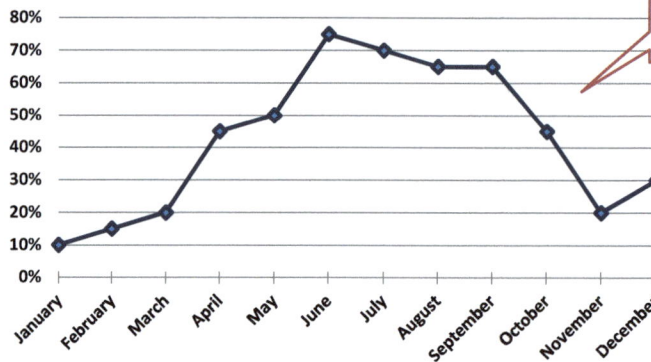

What patterns can you detect?

8

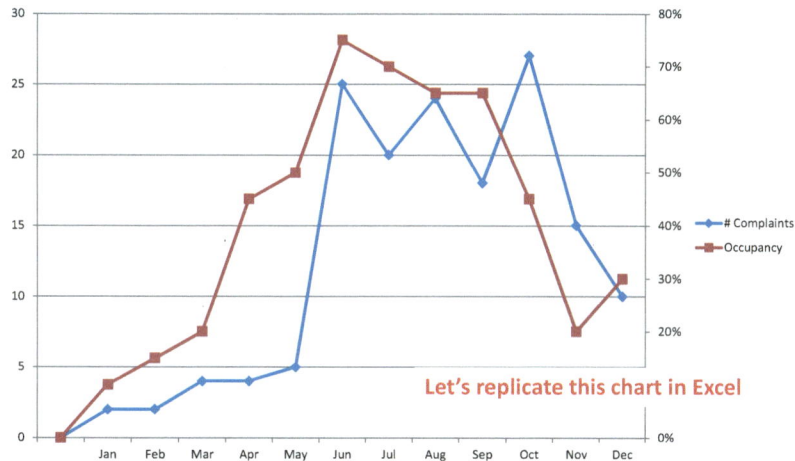

SIR Case Example
Occupancy by Month

Let's replicate this chart in Excel

9

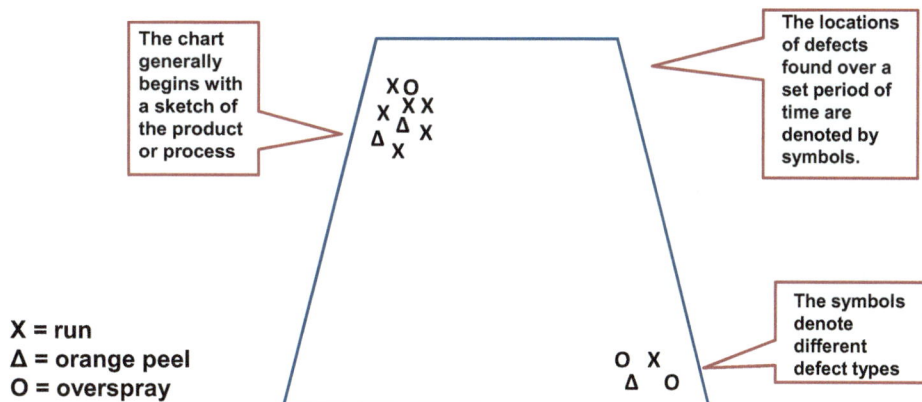

Measles Chart

- A **Measles Chart** shows where defects are physically occurring

The chart generally begins with a sketch of the product or process

The locations of defects found over a set period of time are denoted by symbols.

The symbols denote different defect types

X = run
Δ = orange peel
O = overspray

10

SIR Case Example

Measles Chart

Air Conditioning Complaints
Room Location
May-October

Does the measles chart show any location patterns?

11

Brainstorming Techniques

- The goal is to generate as many ideas as possible in the time allotted
- Don't evaluate each idea as it is mentioned
- Everything gets written on the board as stated!
- Watch for dominant or intimidating personalities
  - Try a round robin approach
  - Consider individual brainstorming time followed by group sharing

12

Affinity Diagrams

- Use an **Affinity Diagram** in a team setting to organize facts, opinions or issues into natural groupings

- "How to wrestle a marshmallow"

3

Affinity Diagram Procedure

1. Gather ideas from a brainstorming or customer feedback session
2. Write each idea on a separate post-it note
3. Place the notes randomly on the board or wall
4. Allow team to SILENTLY move ideas into groupings
5. If a note belongs in more than one group, make a copy
6. When grouping is complete, have team agree on a heading for each group

4

Multi-voting

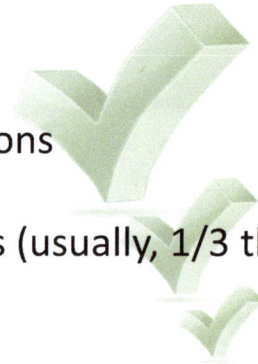

- Used to narrow down a list of ideas or options

- Give each team member a number of votes (usually, 1/3 the number of total items)

- Members cast votes by marking items, or completing a ballot

- Tally votes and identify items at the top and bottom

7

Cause and Effect Diagram

- Used to organize ideas when brainstorming causes of a problem
- Also called a fishbone or Ishikawa diagram

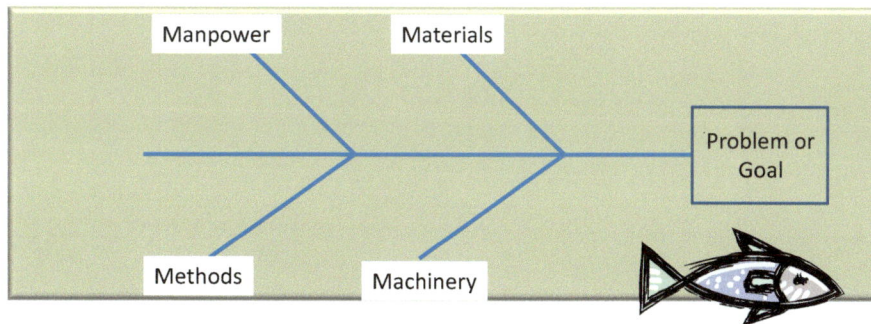

More Graphical Analysis Tools

Define → Measure → Analyze → Improve → Control

1

Frequency Table

Test score raw data:
53, 55, 71, 72, 76, 82, 82, 85, 86, 89, 93, 94, 98, 100

Grade Range	Count
90-100	4
80-89	5
70-79	3
60-69	0
50-59	2

- Categories make "business sense"

- Categories include all data

- There is no overlap between categories

- Each data point fits into one and only one category

- Category ranges are the same size

Do you notice an exception to one of the guidelines here?

2

Histograms

- Graphical display of a frequency table
- Give a sense of the shape of the data distribution

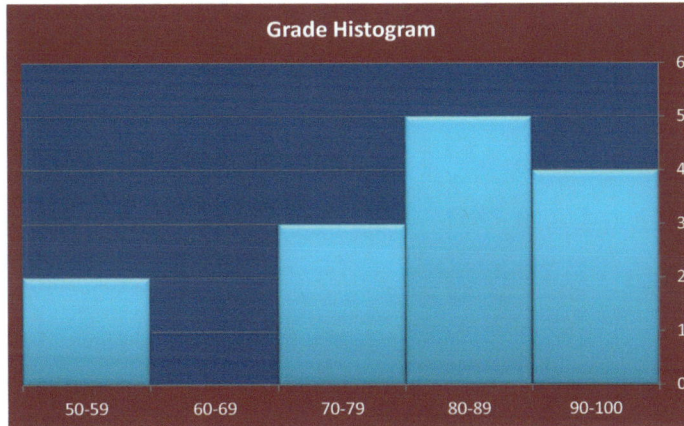

Compare Data Displays

Test score raw data: 53, 55, 71, 72, 76, 82, 82, 85, 86, 89, 93, 94, 98, 100

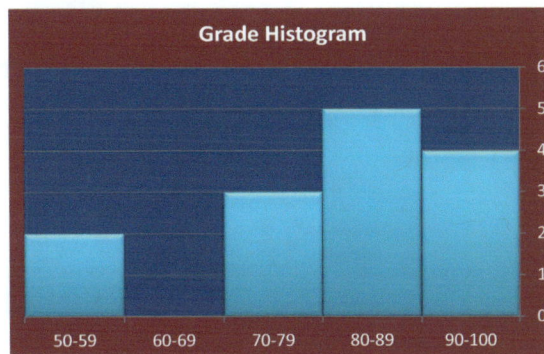

How Many Categories Should We Use?

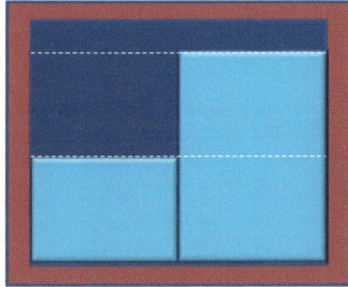

Too few

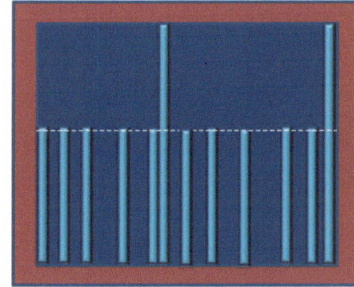

Too many

- Aim for k = 5 to 20 categories

- Can use a rule of thumb: $k \cong \sqrt{n}$

5

Mirasol Wait Times

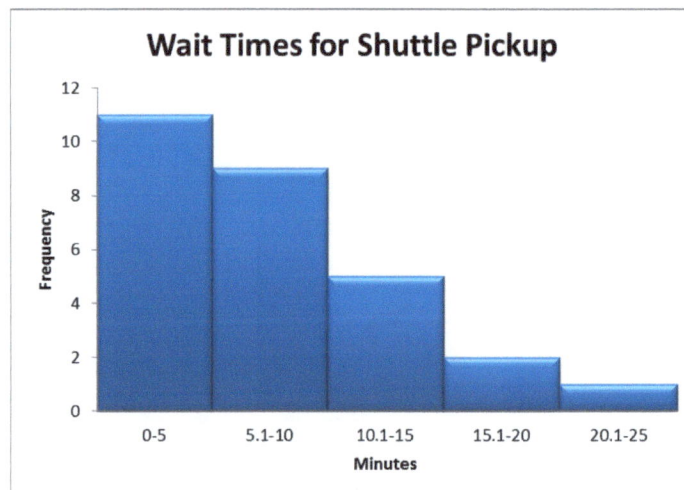

6

Scatter Plots

- A **Scatter Plot** displays the relationship between two variables

    - Height and weight
    - Home price and square footage
    - Room occupancy and revenue

7

SIR Case Example
Avg Length of Stay vs. Revenue

Month	Avg Length of Stay	Revenue
1	1.9	$ 90,500
2	4.6	$ 55,400
3	2.5	$ 100,500
4	2.4	$ 125,400
5	2.2	$ 175,400
6	3.2	$ 490,000
7	1.6	$ 55,800
8	1.7	$ 75,600
9	2.8	$ 130,200
10	3.6	$ 305,100
11	2.5	$ 372,000
12	3.5	$ 697,500
13	3.9	$ 694,400
14	3.6	$ 644,800
15	3.7	$ 487,500
16	1.7	$ 292,950
17	1.5	$ 96,000
18	2.6	$ 195,300

Data was available for the last 18 months

Is this too little data? Should we ignore these results?

Scatter Diagram Analysis

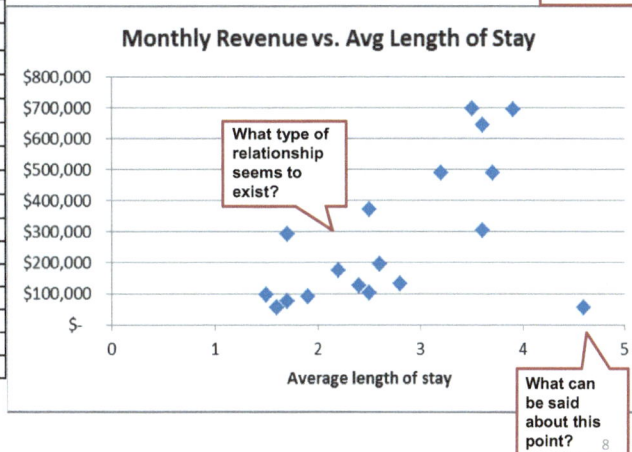

Monthly Revenue vs. Avg Length of Stay

What type of relationship seems to exist?

What can be said about this point?

8

Scatter Diagram Cautions

- Don't confuse correlation with causality
  - Shark attacks and ice cream sales

- Don't extrapolate
  - *"Past performance does not guarantee future results."*

9

Making Sense of Data

Define > Measure > Analyze > Improve > Control

Types of Data

- **Quantitative data** consists of numbers that are counts or measurements

- **Qualitative data** can be separated into different categories based on non-numeric criteria (can't be added or subtracted meaningfully)

© 2017 University Training Partners   2

© 2020 University Training Partners   101

Types of Quantitative Data

- **Continuous data** is measured on a scale that can be infinitely divided
  - Length, weight, time, pressure, volume, etc.

- **Discrete or Attributes data** includes counts, binary data and ordinal data
  - Complaint counts
  - Go/no go checks

3

Statistical Analysis

- **Descriptive Statistics** are used to characterize a data set or population

- **Inferential Statistics** are used to draw conclusions about a population based on analysis of a sample

4

Populations and Samples

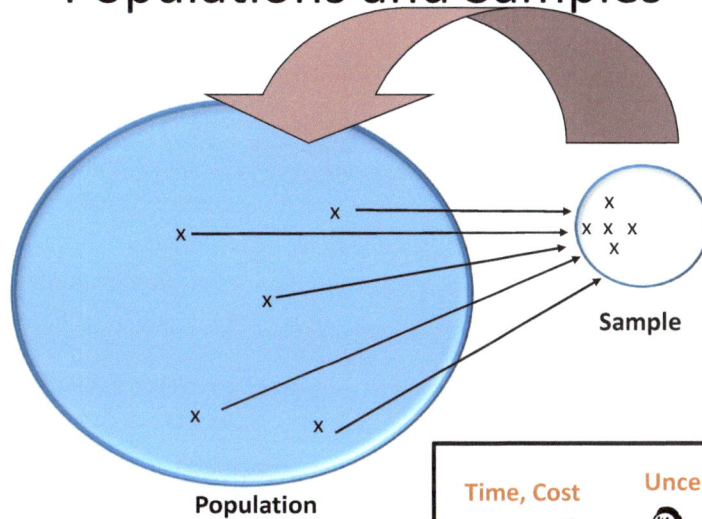

Sample

Population

Time, Cost   Uncertainty

Parameters and Statistics

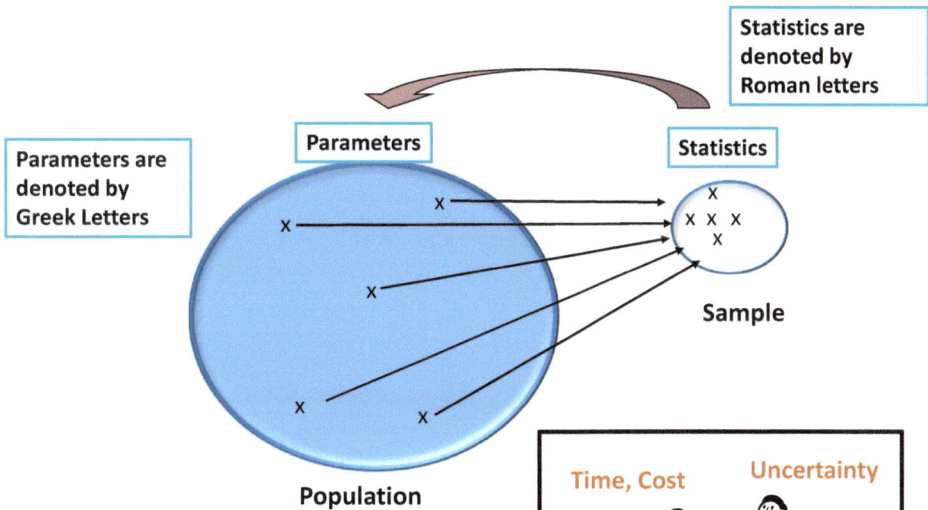

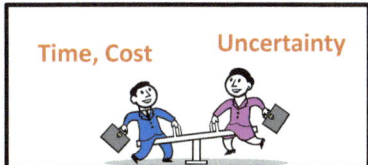

Statistics are denoted by Roman letters

Parameters are denoted by Greek Letters

Parameters

Statistics

Sample

Population

Time, Cost   Uncertainty

Central Tendency and Dispersion

Central Tendency describes the location of a data set.

Dispersion describes the spread of the data values.

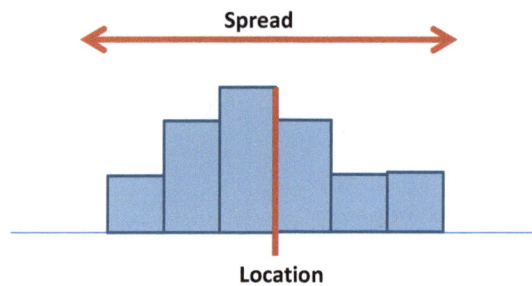

Measures of Central Tendency & Dispersion

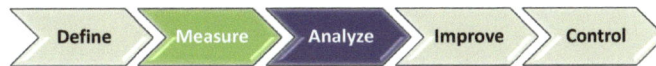

Define · Measure · Analyze · Improve · Control

Module topics

- Introduction to Mean, Median and Mode
- Calculating Mean, Median and Mode
- Features of the Statistics
- Individual Exercise

2

Mean

The **mean** is the arithmetic average of a data set

X bar

Sum up the
data

$$\bar{x} = \frac{\sum x}{n}$$

Divide by the
number of
data points

Median

The **median** is the middle of a data set, $\tilde{x}$

1. Sort the data from low to high

2. If n is odd,

3, 5, ⑦ 12, 13

3. If n is even,

3, 5, ⑦ 12, 13, 15

$$\tilde{x} = (7 + 12) /2 = 9.5$$

Mode

The **mode** is the most frequently occurring value in a data set, $\dot{x}$

1. Sort the data from low to high
2. Mode is the most frequently occurring value

34, 45, 46, 46, 48, 49, 52

5

More on the Mode

23, 25, 26, 28, 29, 30, 32

Find the mode of each data set

65, 67, 67, 69, 70, 70, 83

6

Features of the Statistics

- Mean, $\bar{x}$
  - Uses all data values in the data set
  - Can be influenced by outliers
- Median, $\tilde{x}$
  - Not influenced by outliers
- Mode, $\dot{x}$
  - Not influenced by outliers
  - May not exist
  - May have multiple values

7

Central Tendency and Shape

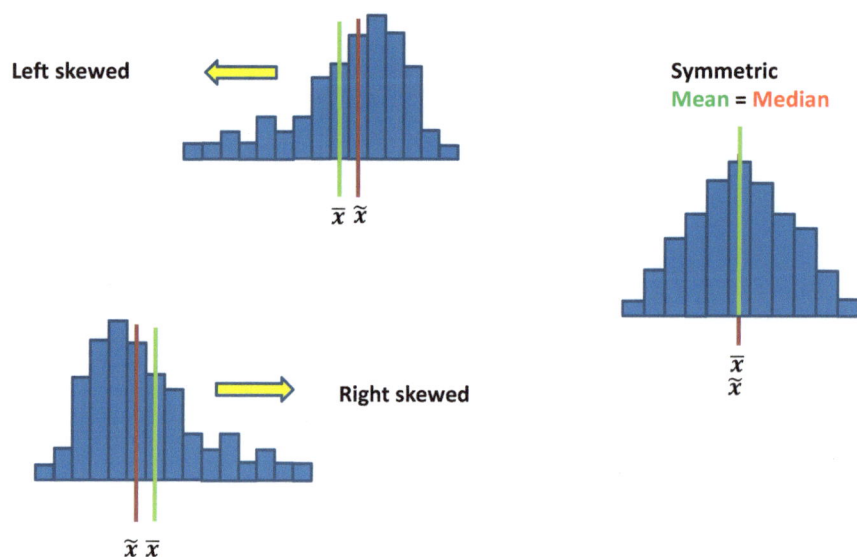

Left skewed

$\bar{x}$ $\tilde{x}$

Right skewed

$\tilde{x}$ $\bar{x}$

Symmetric
Mean = Median

$\bar{x}$
$\tilde{x}$

8

Measures of Dispersion

Define	Measure	Analyze	Improve	Control

Module Topics

- Range & Standard Deviation
- Calculation Class Exercise
- Individual Exercise

10

Range and Standard Deviation

The **range** is the difference between the largest data value and the smallest data value

$$R = (x_{max} - x_{min})$$

The **sample standard deviation** is a measure of the spread of the data

$$s = \sqrt{\frac{\sum(x - \bar{x})^2}{n-1}} \qquad \boxed{s = \sqrt{\frac{\sum x^2 - n\bar{x}^2}{n-1}}}$$

11

Excel Approach

$$= stdev(6,8,8,9,14)$$

Highly recommended!

12

Features of the Range

- The range only depends on the smallest and largest data values in the sample
- Very sensitive to <u>outliers</u>

6, 8, 8, 9, 14

6, 8, 12, 14, 14

6, 13, 13, 13, 14

What are the ranges of these data sets?

What can we say about the standard deviations of these data sets?

6, 13, 13, 13, 100   What is the range?

13

Location and Dispersion

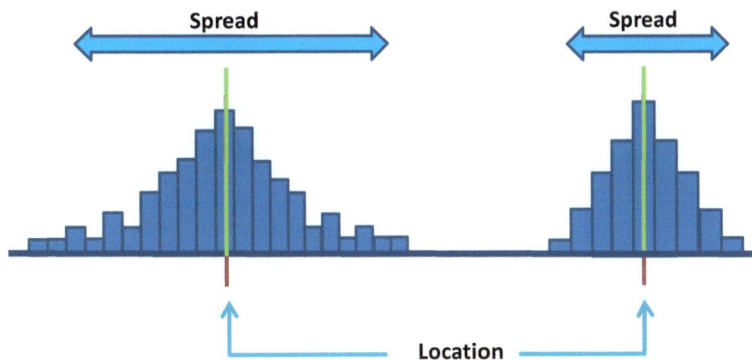

Spread

Spread

Location

Question: What is the shape of each distribution?

14

Normal Distribution

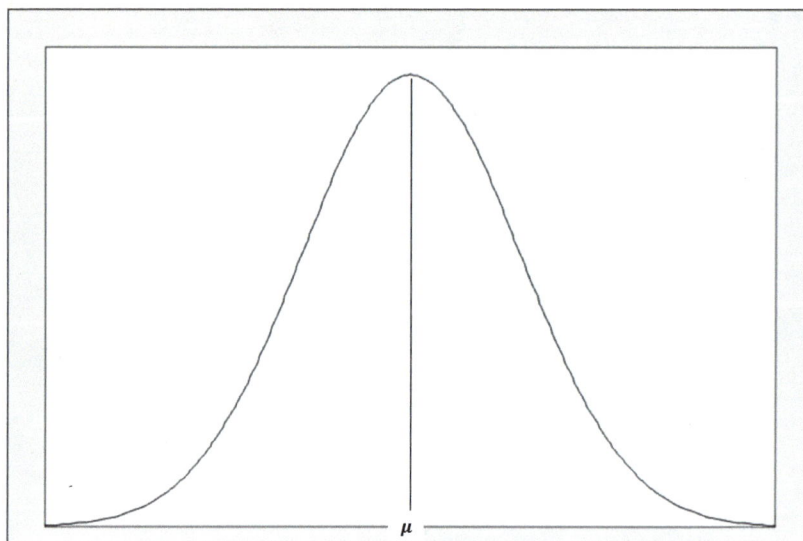

Define Measure Analyze Improve Control

Normal Distribution

μ

Normal Distribution: Standard Deviation

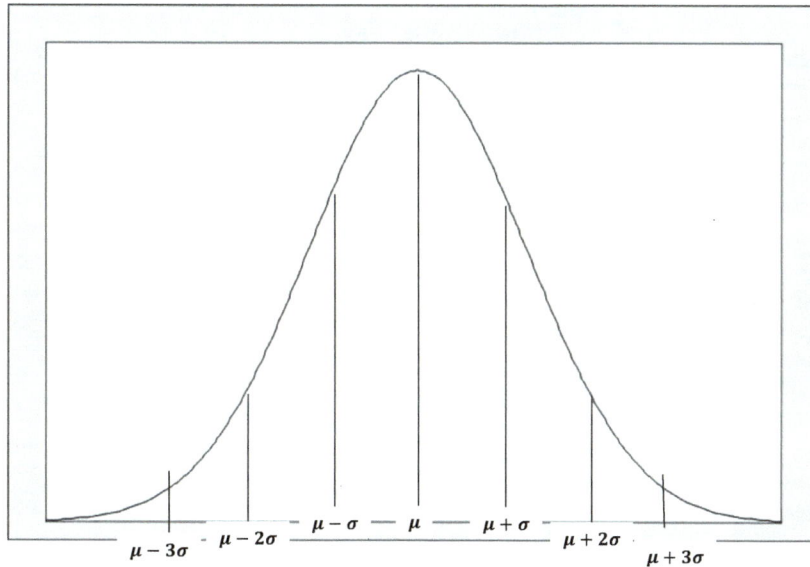

Standard Normal Distribution

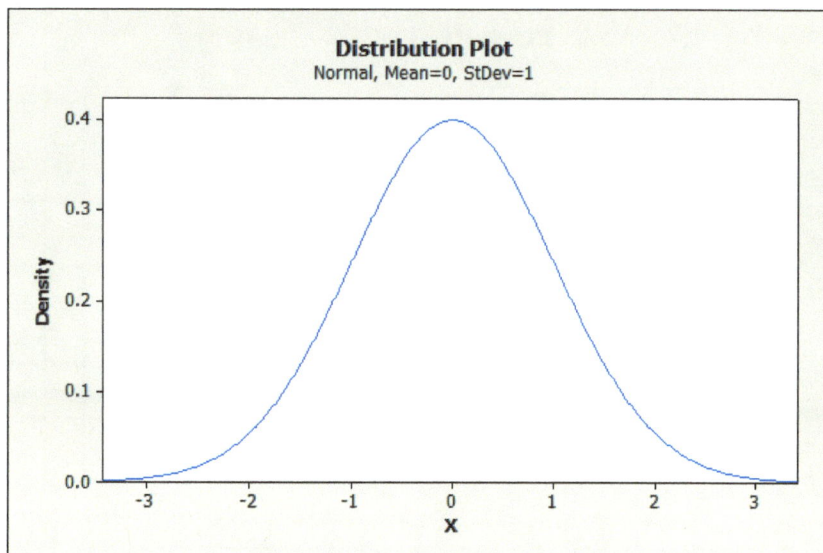

Areas Under the Curve

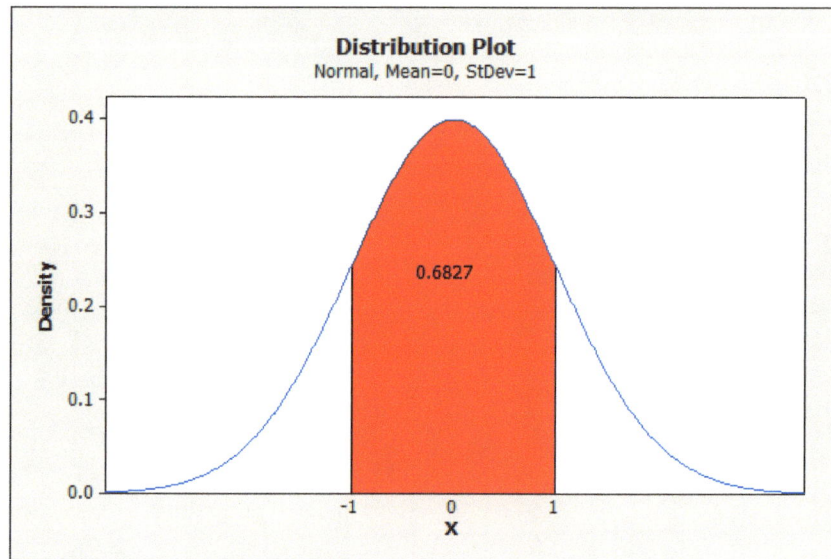

Areas Under the Curve

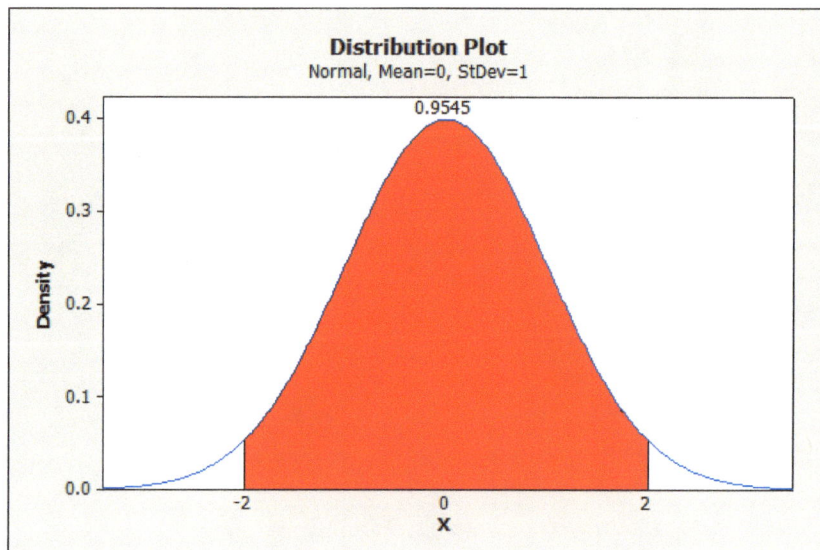

Areas Under the Curve

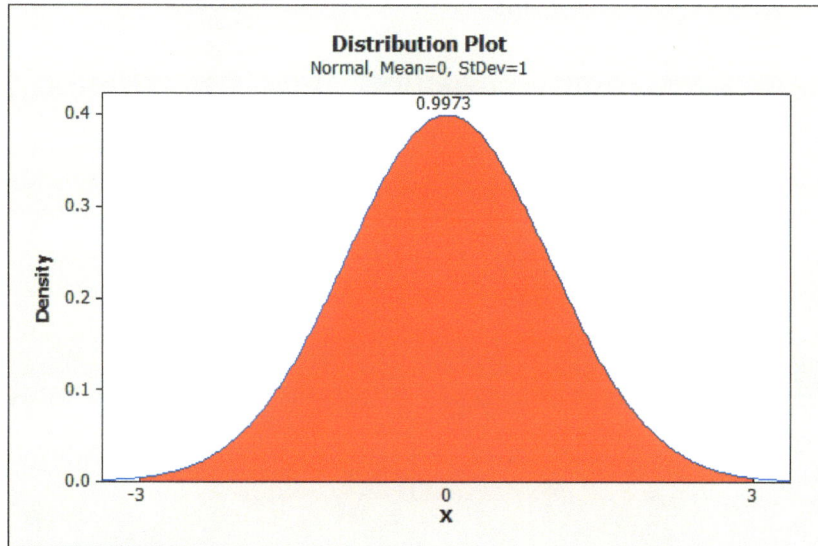

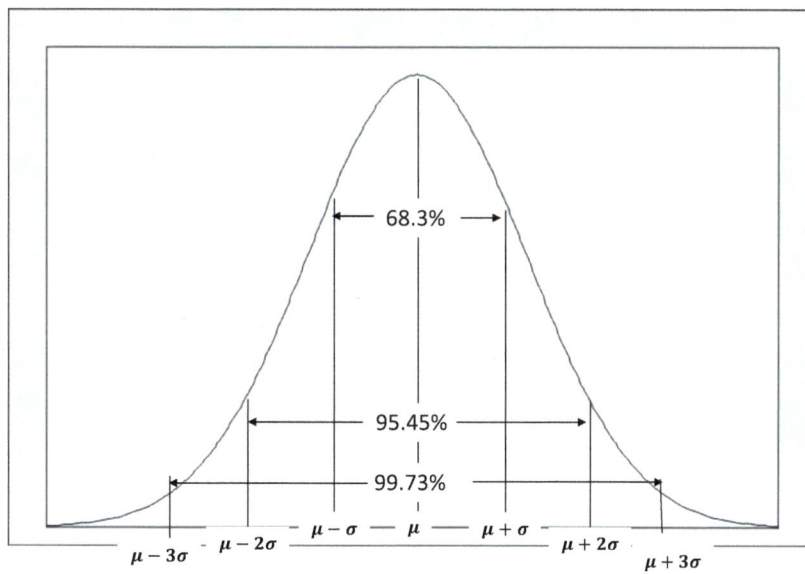

Distribution Plot
Normal, Mean=0, StDev=1

0.9973

Areas Under the Curve

68.3%

95.45%

99.73%

$\mu - 3\sigma$ $\mu - 2\sigma$ $\mu - \sigma$ μ $\mu + \sigma$ $\mu + 2\sigma$ $\mu + 3\sigma$

Z Score & Standard Normal

A **Z score** gives us the relative position of a data value on the normal distribution curve.

Given a data value x from a normal distribution with mean μ and standard deviation σ,

$$z = \frac{(x - \mu)}{\sigma}$$

Areas Under the Standard Normal Curve

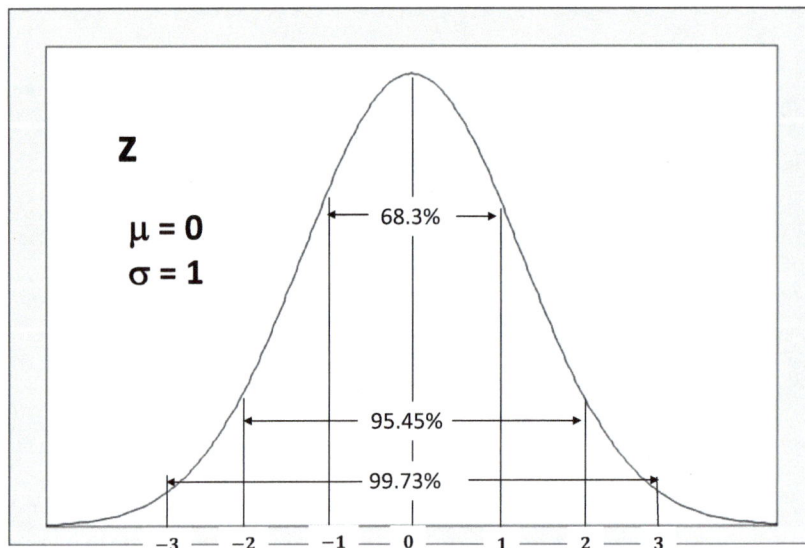

Z

μ = 0
σ = 1

68.3%

95.45%

99.73%

−3   −2   −1   0   1   2   3

How Unusual?

Rule of thumb:

A **Z score** less than -2 or greater than + 2 is considered unusual

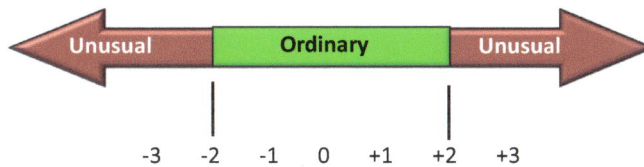

Class Exercise:
What's Your Z Score?

Men's heights follow a normal distribution with mean = 69.0 in and standard deviation = 2.8 in.

Women's heights follow a normal distribution with mean = 63.6 in and standard deviation = 2.5 in.

$$z = \frac{x - \mu}{\sigma}$$

Confidence Intervals

Define ▸ Measure ▸ **Analyze** ▸ Improve ▸ Control

1

How Good is Our Estimate?

- We can estimate the mean using $\overline{X}$

  - This gives us a single point value, but it is very unlikely that the true mean will exactly equal this one point

- A better strategy is to give a range of values that could cover the mean value with a certain probability

  - Takes the uncertainty due to sampling into account

Margin of Error

- We can improve our estimate of the mean by presenting $\overline{X}$ along with a margin of error, E

Size of E depends on:
Confidence level
Sample size
Variation in the data

$$\overline{X} \pm E$$

Confidence level ↑   E ↑

Variation ↑   E ↑

Sample Size ↑   E ↓

Calculating a CI for a Mean, Sigma Known

$$\overline{X} \pm E$$

Based on confidence level

Variability

Sample size

$$E = z\frac{\sigma}{\sqrt{n}}$$

4

"Famous" Z Values

- For a 90% two-sided CI, use Z = 1.645

- For a 95% two-sided CI, use Z = 1.96

- For a 99% two-sided CI, use Z = 2.575

5

Confidence Levels

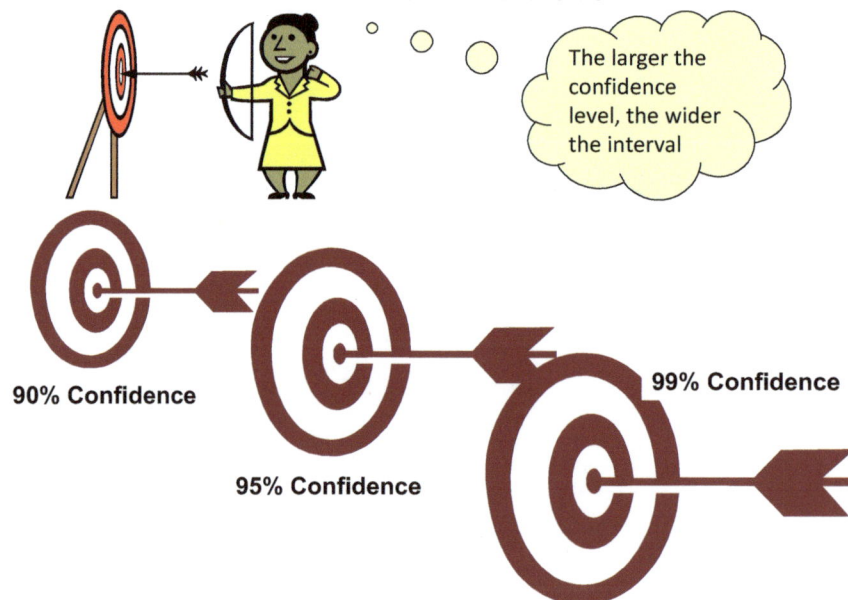

The larger the confidence level, the wider the interval

90% Confidence

95% Confidence

99% Confidence

Calculating a CI for a Mean, Sigma Known

$$\overline{X} \pm E$$

Based on confidence level

$$E = z\frac{\sigma}{\sqrt{n}}$$

Variability

Sample size

Given: $\overline{X}$ = 24.2
σ = 10.0
n = 30

1. Construct a 95 % CI about μ

2. Construct a 99 % CI about μ

7

It's All in the Coverage

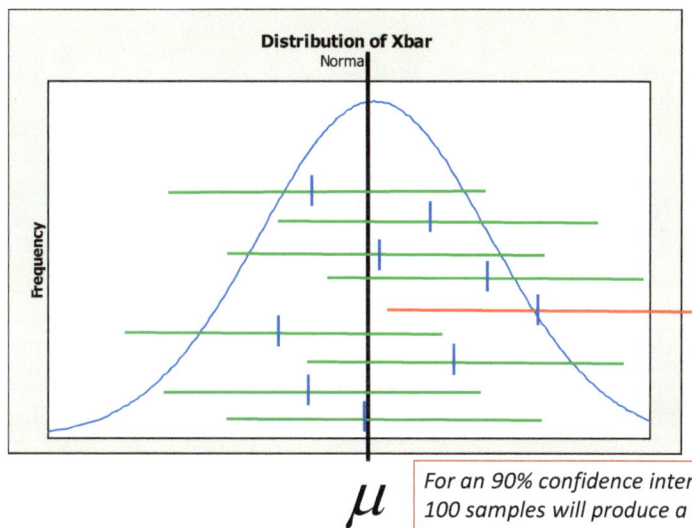

Distribution of Xbar
Normal

Frequency

μ

For an 90% confidence interval, 90 out of 100 samples will produce a confidence interval that covers the true mean.

8

Interpreting a Confidence Interval

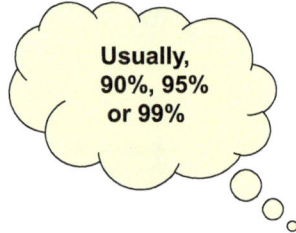

Usually, 90%, 95% or 99%

The mean NEVER falls.

"We are ____ % confident that the true mean is covered by the interval ____, ____."

Six Sigma Concepts

Define > Measure > Analyze > Improve > Control

The 1.5 Sigma Shift

- Dr. Bill Smith at Motorola estimated that in the long run, a well behaved process can vary by as much as 1.5 standard deviations from its mean at any given time.

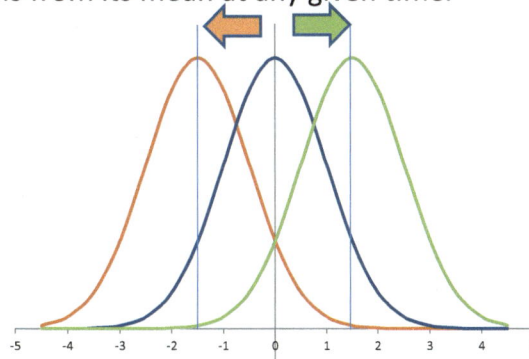

Arbitrary? Perhaps. But the 1.5 sigma shift has become the standard.

1.5 Sigma Shift, Illustrated

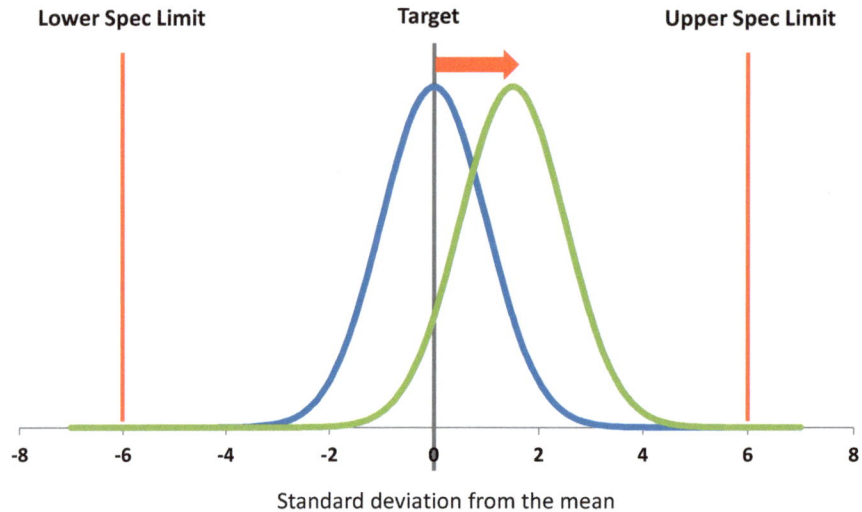

Lower Spec Limit **Target** **Upper Spec Limit**

-8 -6 -4 -2 0 2 4 6 8

Standard deviation from the mean

Short-term and Long-term Quality

- Short-term defect levels do <u>not</u> include the 1.5 sigma shift

- Long-term defect levels include the 1.5 sigma shift

Recall, More Examples

99% Quality = 3.8 sigma	99.99966% Quality = 6 sigma
• 7,011 vehicle breakdowns per day	• 870 breakdowns per year
• 2,441 boating distress calls per month	• 10 distress calls per year
• 3,540 dropped babies per week	• 63 dropped babies per year
• 1,452 surgical mistakes per day	• 180 surgical mistakes per year

With 1.5 sigma shift

5

Sigma Level Examples

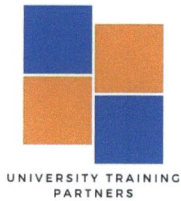

# defectives	2
sample size	100
% Defective	2%
% Yield	98%
DPMO	20,000

NO 1.5 sigma shift!

Sigma	DPMO	Yield (%)	Sigma	DPMO	Yield (%)
6.0	0.0	100.00000	3.0	2,700	99.73
5.9	0.0	100.00000	2.9	3,732	99.63
5.8	0.0	100.00000	2.8	5,110	99.49
5.7	0.0	100.00000	2.7	6,934	99.31
5.6	0.0	100.00000	2.6	9,322	99.07
5.5	0.0	100.00000	2.5	12,419	98.76
5.4	0.1	99.99999	2.4	16,395	98.36
5.3	0.1	99.99999	2.3	21,448	97.86
5.2	0.2	99.99998	2.2	27,807	97.22
5.1	0.3	99.99997	2.1	35,729	96.43
5.0	0.6	99.99994	2.0	45,500	95.45
4.9	1.0	99.99990	1.9	57,433	94.26
4.8	1.6	99.99984	1.8	71,861	92.81
4.7	2.6	99.99974	1.7	89,131	91.09
4.6	4.2	99.99958	1.6	109,599	89.04
4.5	6.8	99.99932	1.5	133,614	86.64
4.4	10.8	99.99892	1.4	161,513	83.85
4.3	17.1	99.99829	1.3	193,601	80.64
4.2	26.7	99.99733	1.2	230,139	76.99
4.1	41.3	99.99587	1.1	271,332	72.87
4.0	63.3	99.99367	1.0	317,311	68.3
3.9	96.2	99.99038	0.9	368,120	63.2
3.8	144.7	99.98553	0.8	423,711	57.6
3.7	215.6	99.97844	0.7	483,927	51.6
3.6	318.2	99.96818	0.6	548,506	45.1
3.5	465.3	99.95347	0.5	617,075	38.3
3.4	673.9	99.93261	0.4	689,157	31.1
3.3	966.8	99.90332	0.3	764,177	23.6
3.2	1374.3	99.86257	0.2	841,481	15.9
3.1	1935.2	99.80648	0.1	920,344	8.0

Includes the 1.5 sigma shift

Sigma	DPMO	Yield (%)	Sigma	DPMO	Yield (%)
6.0	3.4	99.99966	3.0	66,811	93.32
5.9	5.4	99.99946	2.9	80,762	91.92
5.8	8.5	99.99915	2.8	96,809	90.32
5.7	13	99.99867	2.7	115,083	88.49
5.6	21	99.99793	2.6	135,687	86.43
5.5	32	99.99683	2.5	158,687	84.13
5.4	48	99.99519	2.4	184,108	81.59
5.3	72	99.99277	2.3	211,928	78.81
5.2	108	99.98922	2.2	242,071	75.79
5.1	159	99.98409	2.1	274,412	72.56
5.0	233	99.9767	2.0	308,770	69.12
4.9	337	99.9663	1.9	344,915	65.51
4.8	483	99.9517	1.8	382,572	61.74
4.7	687	99.9313	1.7	421,427	57.86
4.6	968	99.9032	1.6	461,140	53.89
4.5	1,350	99.8650	1.5	501,350	49.87
4.4	1,866	99.8134	1.4	541,694	45.83
4.3	2,555	99.7445	1.3	581,815	41.82
4.2	3,467	99.6533	1.2	621,378	37.86
4.1	4,661	99.5339	1.1	660,083	33.99
4.0	6,210	99.379	1.0	697,672	30.23
3.9	8,198	99.180	0.9	733,944	26.61
3.8	10,724	98.928	0.8	768,760	23.12
3.7	13,904	98.610	0.7	802,048	19.80
3.6	17,865	98.214	0.6	833,804	16.62
3.5	22,750	97.725	0.5	864,095	13.59
3.4	28,717	97.128	0.4	893,050	10.69
3.3	35,931	96.407	0.3	920,861	7.91
3.2	44,567	95.543	0.2	947,765	5.22
3.1	54,801	94.52	0.1	974,043	2.6

Calculate the sigma level for this process. What are the corresponding short-term and long-term DPMO values?

s = 10.0

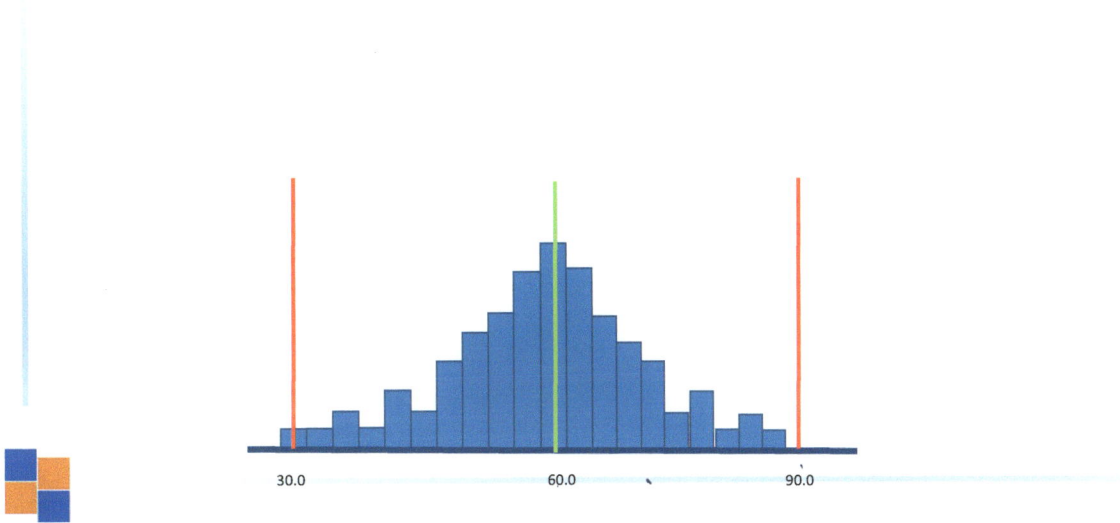

Shuttle wait times

Based on customer feedback, Mirasol has set the upper limit for wait time to 15 minutes, with a target of 9 minutes and a lower limit of 3 minutes. Waiting a few minutes for the shuttle gives arriving guests a chance to stop at the water sports kiosk adjacent to the dock. The kiosk has information on dolphin cruise schedules, deep sea fishing excursions and rentals on jet skis, kite boards, kayaks and boats.

• Draw in the customer requirements: target and upper and lower limits
• Comment on the shape, center and variation of the data. Is it skewed? Centered? What is the spread?
• Determine the % of the data points that are outside the limits. What is the sigma level, and the short and long-term yield?

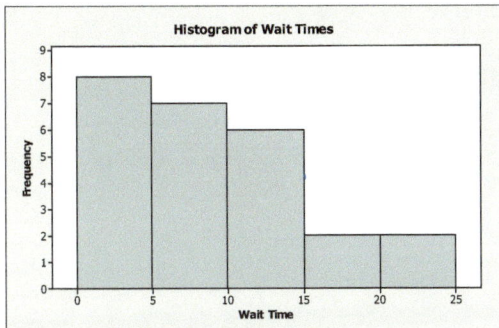

Wait Times for Shuttle Pickup at Ferry Dock

3.5	5.5	12.0	8.0	15.0
20.0	14.0	18.0	9.5	12.5
1.5	4.0	5.0	3.0	4.0
0.0	7.5	10.5	3.0	25.0
9.0	10.0	7.0	4.0	10.0
4.0	4.5	9.5		

Laundry Facility Water Usage

Water and wastewater disposal comprise half of the total costs of a commercial laundry facility. The measure of efficiency for commercial laundry facilities is measured in gallons of water per pound of fabric. This metric was recorded for 25 consecutive wash cycles, with the results shown below. The service contractor claims that the current wash cycle design should have a target of 4.8 gal/lb with lower limit = 4.5 gal/lb and an upper limit = 5.1 gal/lb.

• Draw in the customer requirements on the histogram: target and upper and lower limits
• Comment on the shape, center and variation of the data. Is it skewed? Centered? What is the spread?
• Determine the % of the data points that are outside the limits. What is the sigma level, and the short and long-term DPMO?
• What should be done to improve the sigma level?

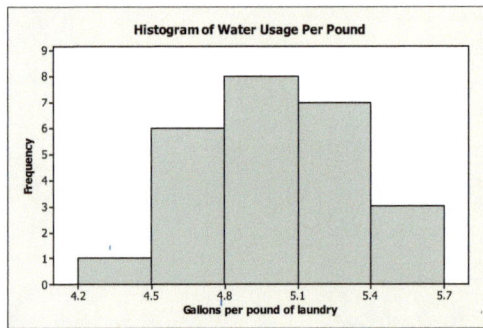

Histogram of Water Usage Per Pound

N	25
Sample Mean	5.0
Sample Standard Deviation	0.3
Minimum	4.3
Maximum	5.6

Target	4.8
Lower specification	4.5
Upper specification	5.1

Process Capability

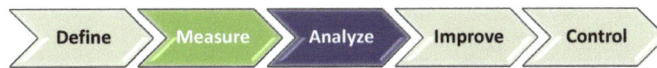

Define Measure Analyze Improve Control

Capability Indices

- **Capability** is a measure of the inherent uniformity of the process

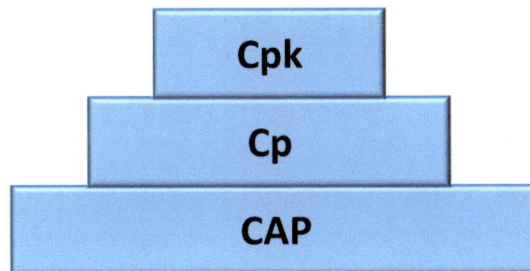

Cpk

Cp

CAP

Capability Requirements

- The process must be predictable (in-control)

- The process metric must be normally distributed

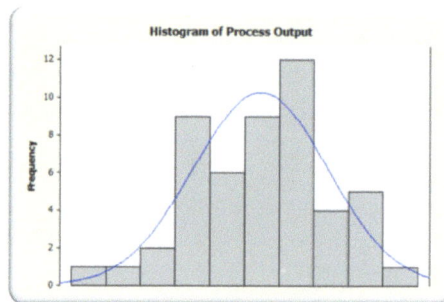

CAP

- **CAP** is also known as the **natural tolerance**
- CAP is a measure of the total spread of the process

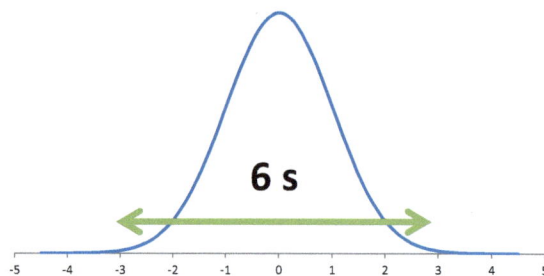

This is not "Six Sigma"

$$\widehat{CAP} = 6\ s$$

6 s

Cp

Cp compares the specification width to the natural tolerance of the process

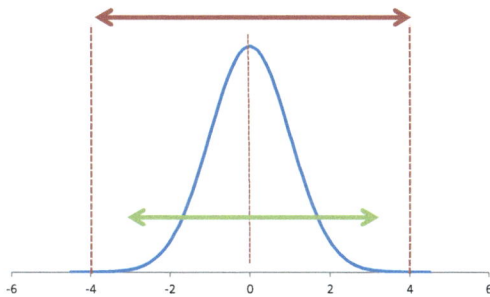

$$\hat{C}_p = \frac{(USL - LSL)}{6s}$$

5

Cp Examples

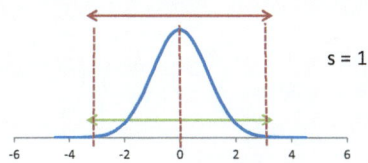

$s = 1$

$$\hat{C}_p = \frac{(USL - LSL)}{6s} = 1$$

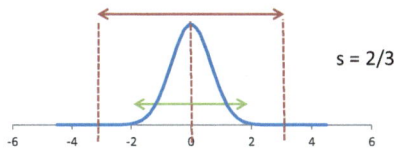

$s = 2/3$

$$\hat{C}_p = 1.5$$

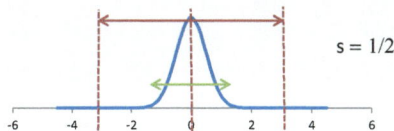

$s = 1/2$

$$\hat{C}_p = 2.0$$

6

Cp

Compare the Cp values of these processes

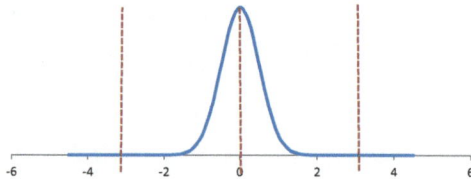

$$\hat{C}_p = 2.0$$

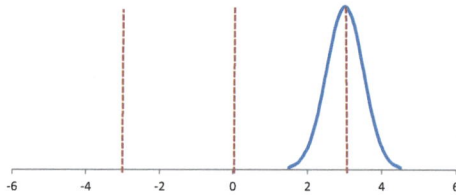

$$\hat{C}_p = 2.0$$

Cpk

- **Cpk** compares both the spread and the location of the process to the specifications

$$\hat{C}_{pl} = \frac{(\mu - LSL)}{3s} \qquad \hat{C}_{pu} = \frac{(USL - \mu)}{3s}$$

$$\hat{C}_{pk} = min(\hat{C}_{pl}, \hat{C}_{pu})$$

Cpk

For a centered process, Cp = Cpk

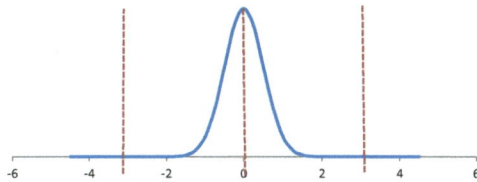

$$\hat{C}_{pk} = 2.0$$
$$\hat{C}_{p} = 2.0$$

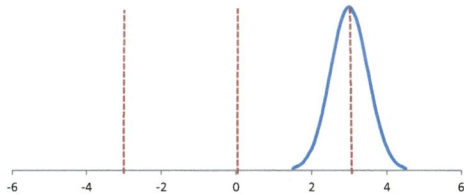

$$\hat{C}_{pk} = 0$$
$$\hat{C}_{p} = 2.0$$

9

Capability and Sigma Level

Sigma	Cp	DPMO	Yield (%)
6.0	2.00	0.0	100.0000
5.0	1.67	0.6	99.9999
4.0	1.33	63.3	99.9937
3.0	1.00	2,700	99.7300
2.0	0.67	45,500	95.4500
1.0	0.33	317,311	68.2689

Does not include the 1.5 sigma shift

10

Class Exercise: Capability Calculation

- Find the estimated CAP, Cp and Cpk for a normally distributed, stable process variable with mean = 54.5 and standard deviation = 2.0

- Customer requirements (specifications) have a target = 56.0, lower limit = 48.0 and upper limit = 64.0

- What should be done to improve the process?

11

Group Exercise: Capability Calculations

- Revisit the Sigma Calculation exercise, and calculate the estimated CAP, Cp and Cpk for each scenario, if appropriate.

12

Capability Examples

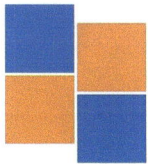

Shuttle wait times

Based on customer feedback, Mirasol has set the upper limit for wait time at the ferry dock to 15 minutes, with a target of 9 minutes and a lower limit of 3 minutes. Waiting a few minutes for the shuttle gives arriving guests a chance to stop at the water sports kiosk adjacent to the dock. The kiosk has information on dolphin cruise schedules, deep sea fishing excursions and rentals on jet skis, kite boards, kayaks and boats.

- Draw in the customer requirements: target and upper and lower limits
- Comment on the shape, center and variation of the data. Is it skewed? Centered? What is the spread?
- Assuming stability, can capability be calculated here? If so, estimate CAP, Cp and Cpk.

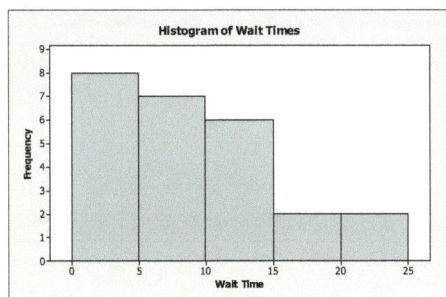

Histogram of Wait Times

Laundry Facility Water Usage

Water and wastewater disposal comprise half of the total costs of a commercial laundry facility. The measure of efficiency for commercial laundry facilities is measured in gallons of water per pound of fabric. This metric was recorded for 25 consecutive wash cycles, with the results shown below. The service contractor claims that the current wash cycle design should have a target of 4.8 gal/lb with lower limit = 4.5 gal/lb and an upper limit = 5.1 gal/lb.

- Draw in the customer requirements on the histogram: target and upper and lower limits
- Comment on the shape, center and variation of the data. Is it skewed? Centered? What is the spread?
- Assuming stability, can capability be calculated here? If so, estimate CAP, Cp and Cpk.

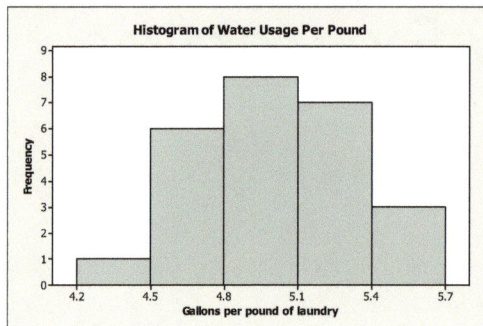

Histogram of Water Usage Per Pound

N	25	Target	4.8
Sample Mean	5.0	Lower specification	4.5
Sample Standard Deviation	0.3	Upper specification	5.1
Minimum	4.3		
Maximum	5.6		

Control Charts

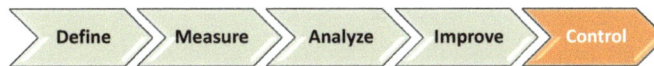

Define > Measure > Analyze > Improve > Control

Module Topics

- Variation Exercise
- Purpose of Control Charts
- Types of Data
- Variables Charts
  - X and MR
  - X-Bar and R
- Attributes Charts
  - NP, P
  - C, U
- Out of Control vs. Out of Spec

Variation Exercise

Everything varies!

1. Random or common cause variation
2. Special cause variation

Name Game:

Write your name 5 times on the paper provided.

Now write your name with your other hand.

3

Purpose of Control Charts

- A graphical way to track a process variable over time
- A statistical method to determine whether the process has changed
  - Is variation due to common causes?
  - Is variation due to special causes?

4

Out-of-Control

- Control charts can signal an out-of-control condition
- Indicates that a special cause of variation is present
- Out-of-control implies that the process is not predictable

Recall: How Unusual?

Rule of thumb:

A **Z score** less than -2 or greater than + 2 is considered unusual

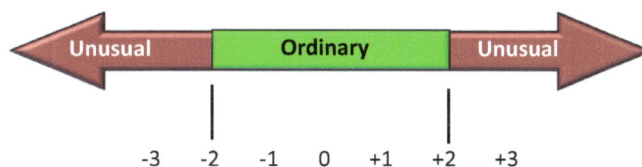

Unusual	Ordinary	Unusual

-3   -2   -1   0   +1   +2   +3

For Control Charts...

Rule of thumb:

A **plotted point** less than -3σ or greater than + 3σ is considered unusual, or **out of control**

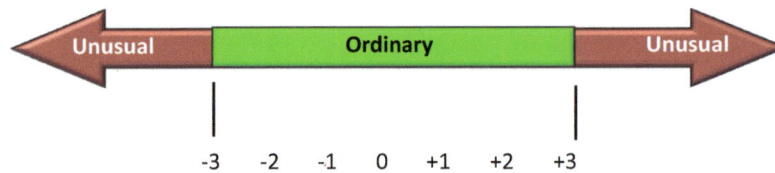

This practice prevents "false alarms" and unnecessary line stoppages.

7

Recall Types of Quantitative Data

- **Continuous or Variables data** is measured on a scale that can be infinitely divided
  - Length, weight, time, pressure, volume, etc.

- **Discrete or Attribute data** includes counts, binary data
  - Defect counts per unit
  - Counts of defective units

8

Types of Charts by Data Type

Variables

| X-MR |
| X-bar & R |
| X-bar & S |
| EWMA |

Attribute

| NP |
| P |
| C |
| U |

9

Variables

| X-MR |
| X-bar & R |
| X-bar & S |
| EWMA |

10

Variables Charts: X and MR

- **X is a normally distributed variable**
- **The moving range is the difference between two consecutive points**
- **Subgroup size = 1**

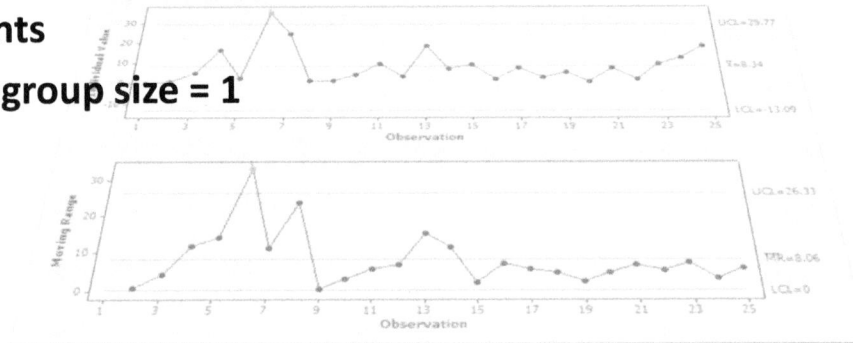

Variables Charts: X and MR

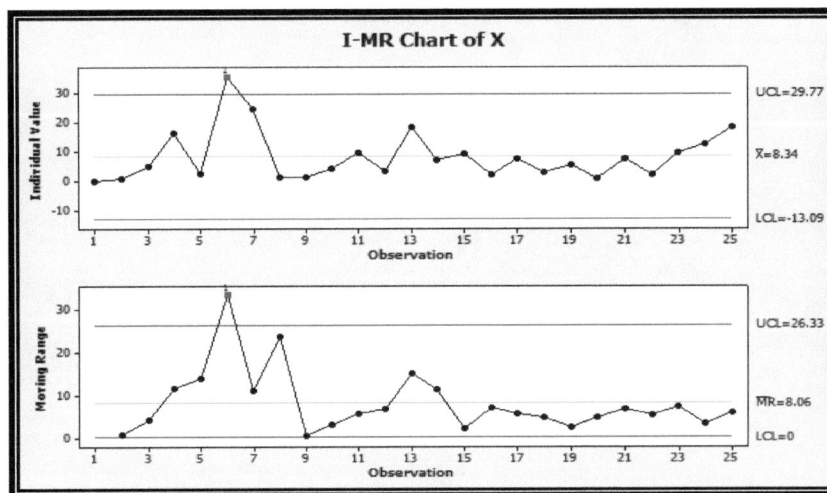

Variables Charts: X-Bar and R Chart

- **Use subgroup sizes from 3 to 6**
- **X-bar is the average of the subgroup**
  Tracks the location of the process
- **R is the range of the subgroup**
  Tracks the variability within the subgroups

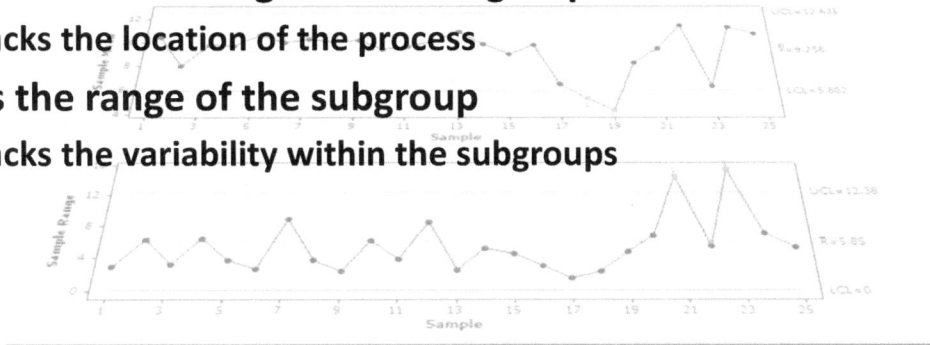

Variables Charts: X-bar and R

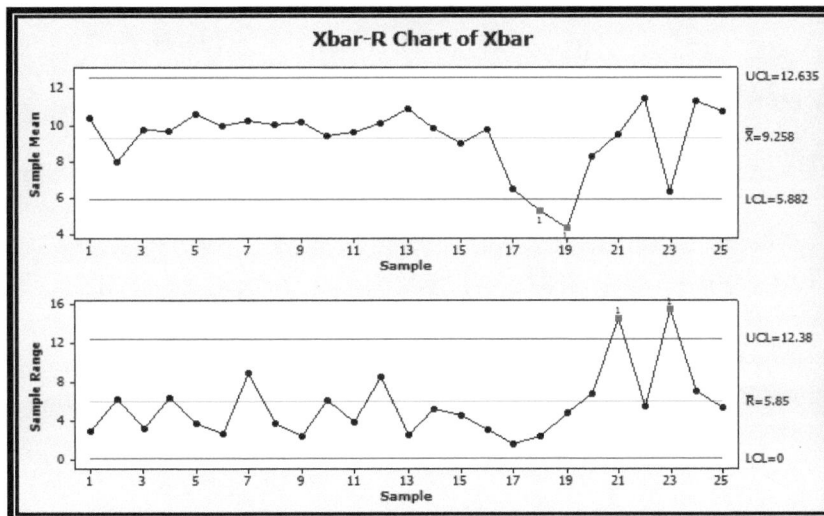

Attribute

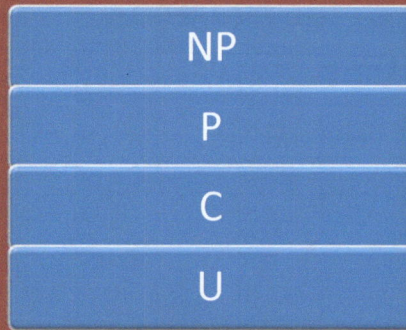

NP

P

C

U

Defectives vs. Defects

- A **defective** unit is found using a pass/fail, go/no go type of test

- There can be many **defects** in a single unit

Attributes Charts: NP

- Products are classified as **defective** or non-defective (go/ no go)
- NP is the number of **defective** units found in the sample
- Sample size is constant

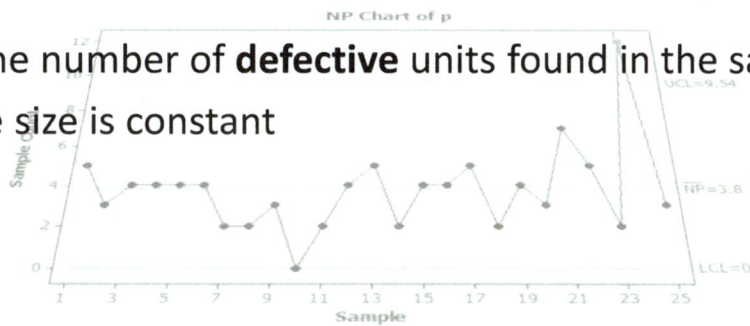

Attributes Charts: NP

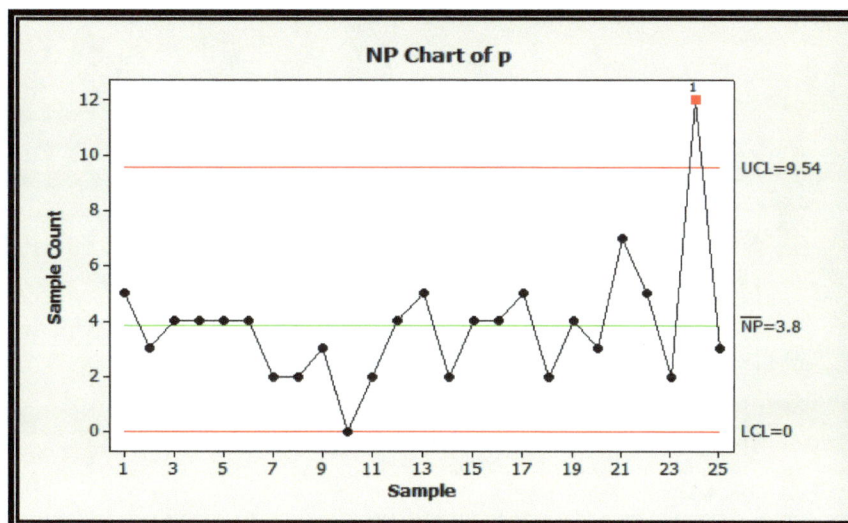

Attributes Charts: P

- Products are classified as **defective** or non-defective (go/no go)
- P is the percent **defective** in the sample
- Sample size may not be constant
- Control limits change width according to sample size

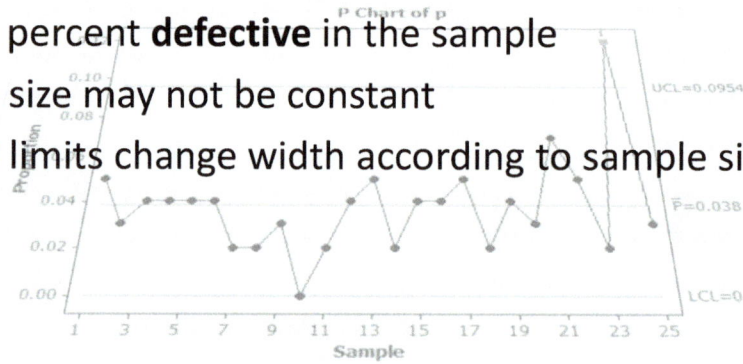

Attributes Charts: P

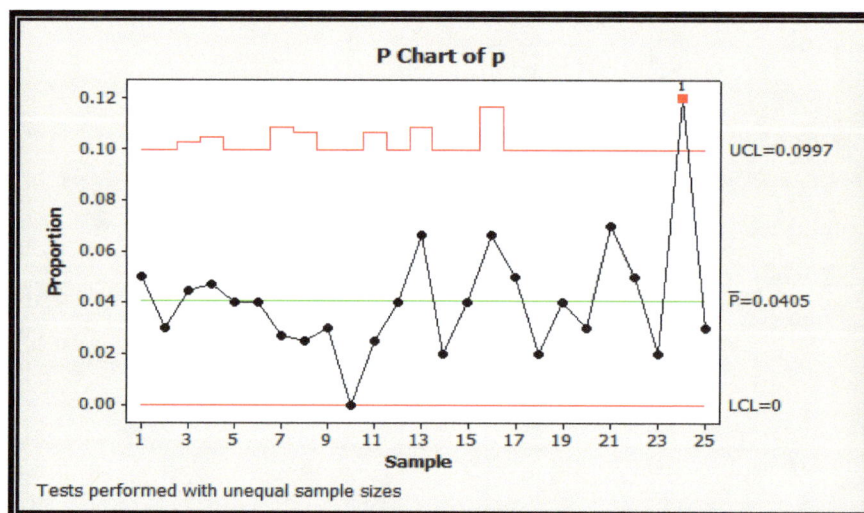

P Chart of p

Tests performed with unequal sample sizes

Attributes Charts: C

- Products are examined and the number of **defects** is recorded
- C is the number of **defects** found in a sample
- Sample size is constant

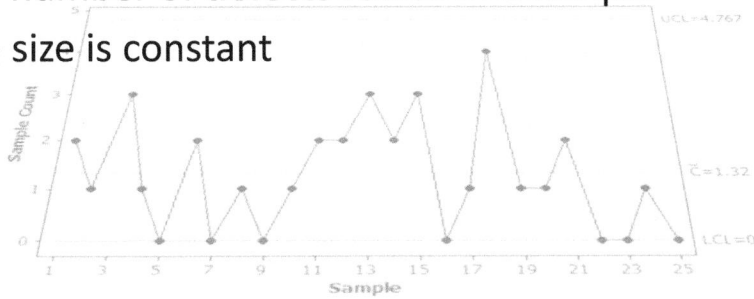

Attributes Charts: C

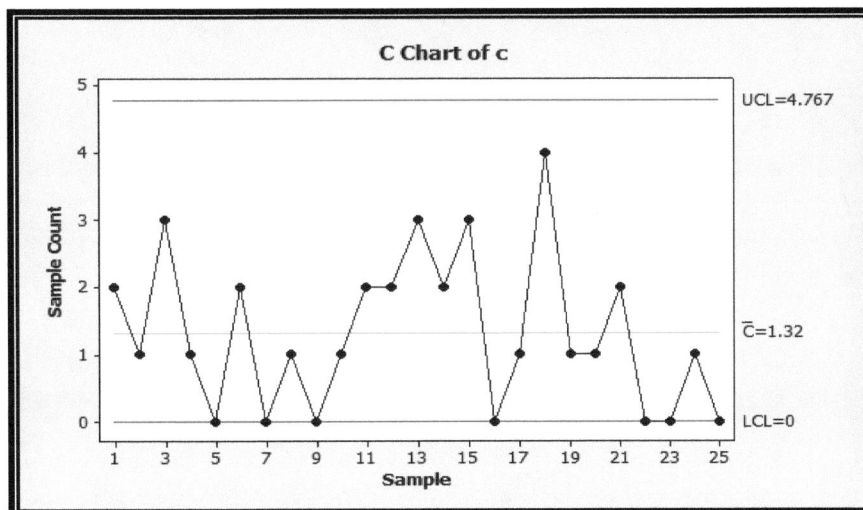

21

22

Attributes Charts: U

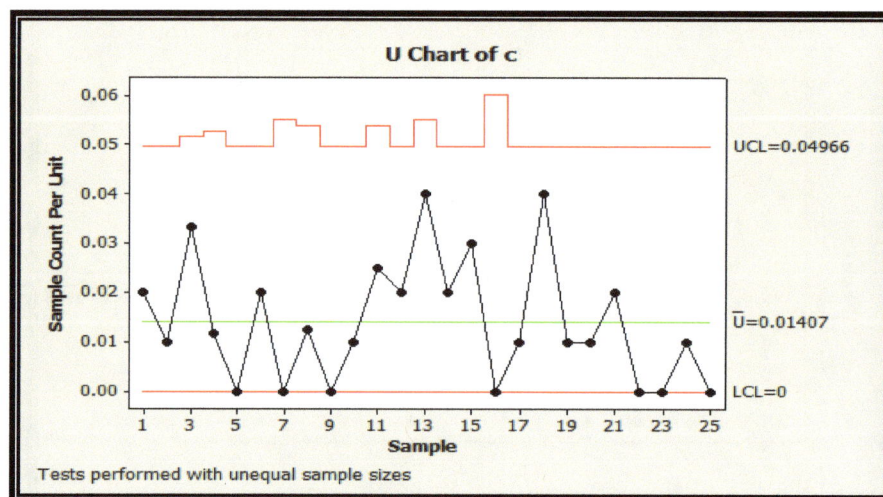

- Products are examined and the number of **defects** is recorded
- U is the average number of **defects** per unit
- Sample size may not be constant
- Control limits change width according to sample size

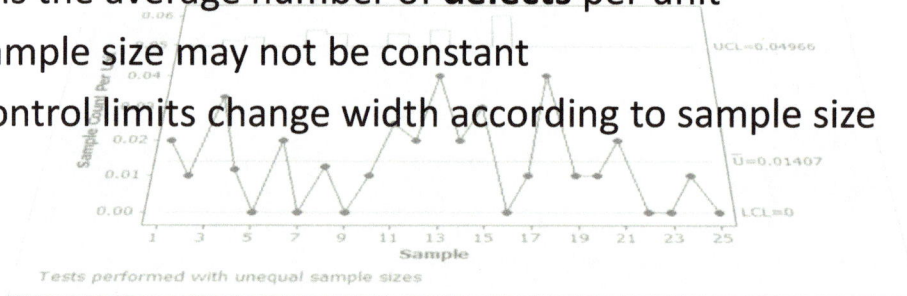

Attributes Charts: U

Which Chart?

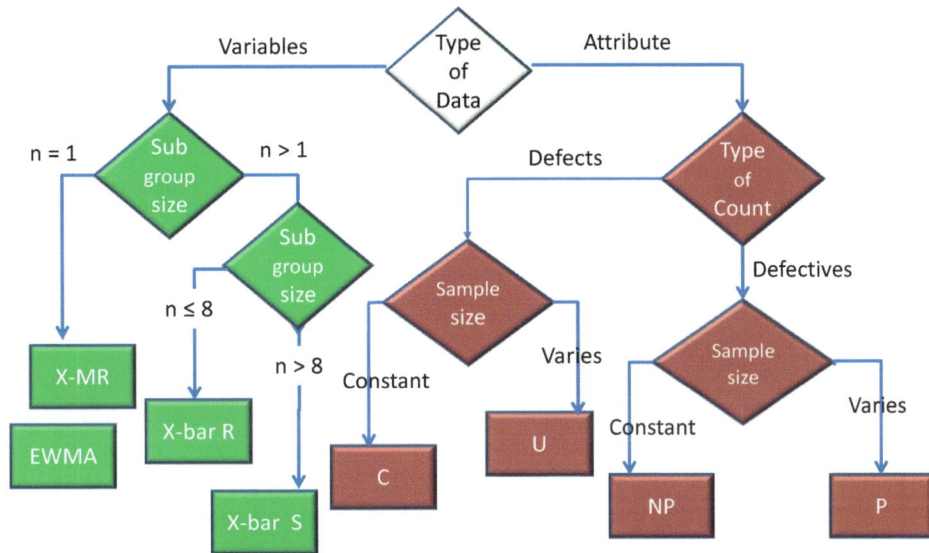

25

Out of Control vs. Out of Spec

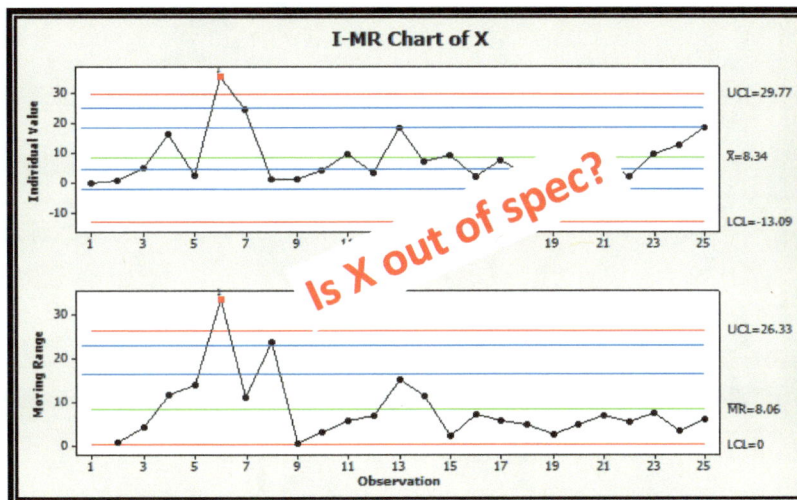

26

NP Chart: Out of Spec?

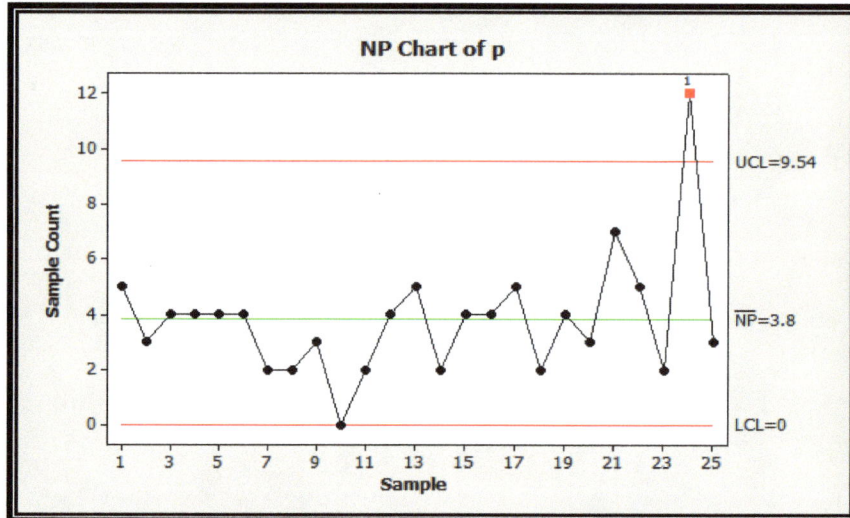

C Chart: Out of Spec?

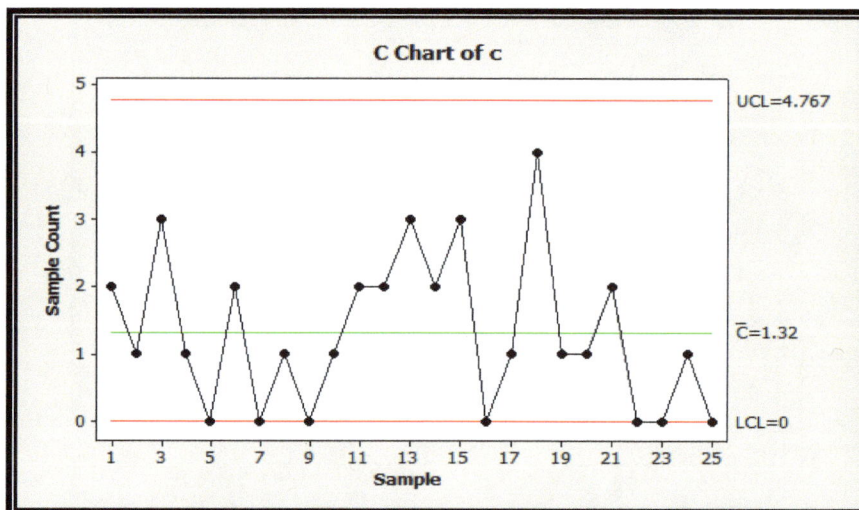

Which Chart?

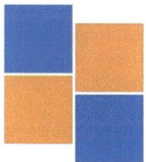

Types of Charts by Data Type

Variables	Attribute
X-MR	NP
X-bar & R	P
X-bar & S	C
EWMA	U

Defectives vs. Defects

- A **defective** unit is found using a pass/fail, go/no go type of test

- There can be many **defects** in a single unit

Which Chart?

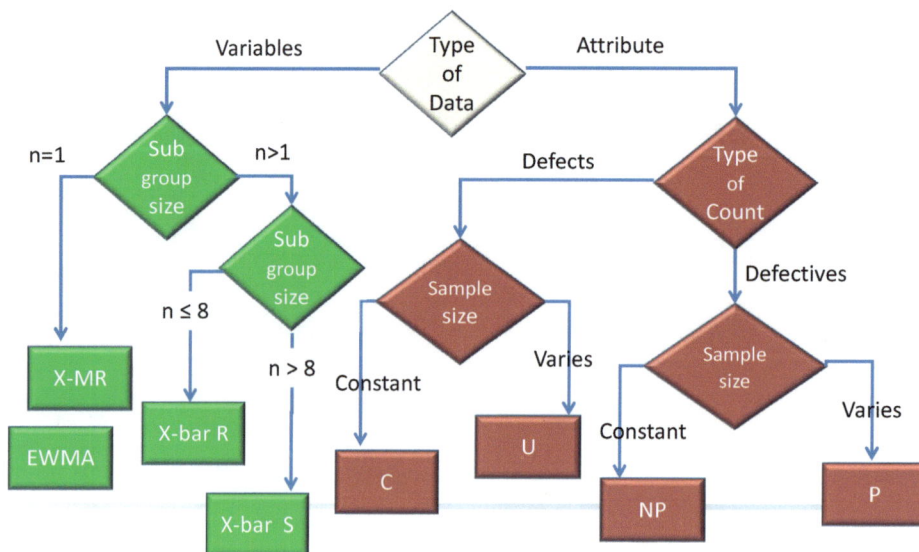

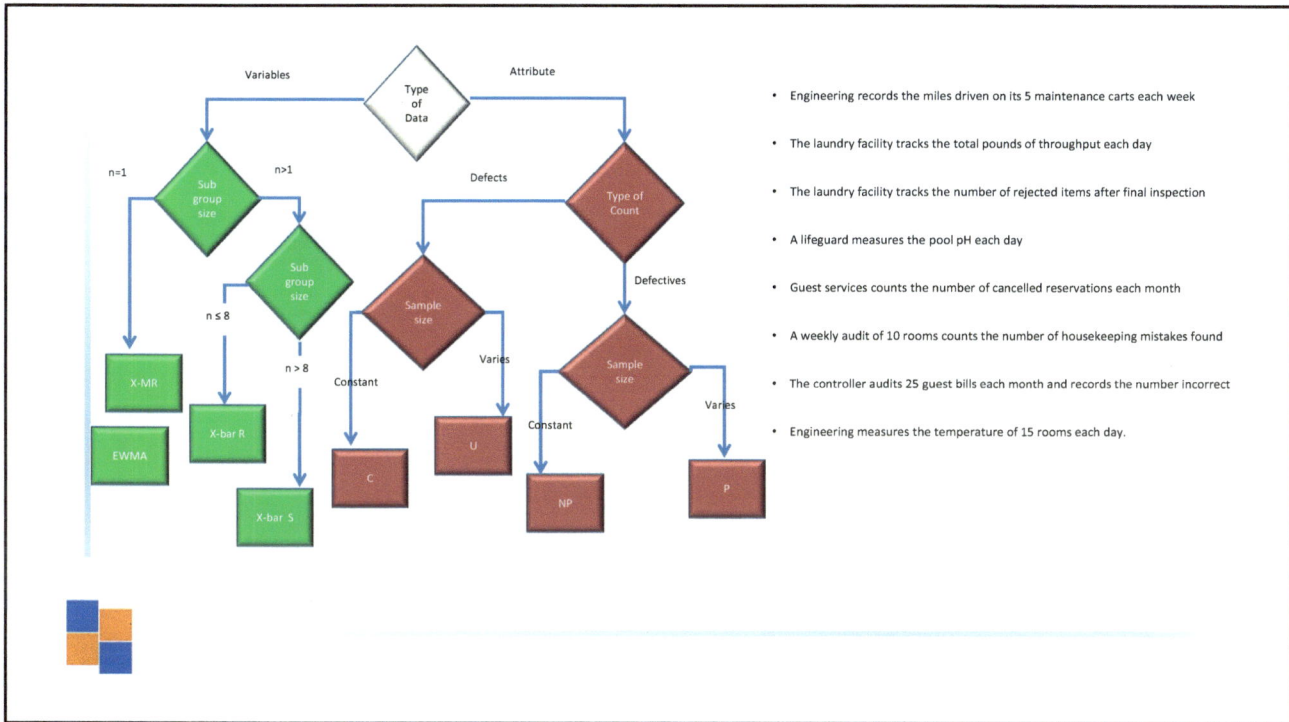

- Engineering records the miles driven on its 5 maintenance carts each week

- The laundry facility tracks the total pounds of throughput each day

- The laundry facility tracks the number of rejected items after final inspection

- A lifeguard measures the pool pH each day

- Guest services counts the number of cancelled reservations each month

- A weekly audit of 10 rooms counts the number of housekeeping mistakes found

- The controller audits 25 guest bills each month and records the number incorrect

- Engineering measures the temperature of 15 rooms each day.

SPC Chart Patterns

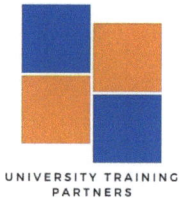

Control Chart Zones

Control Plans

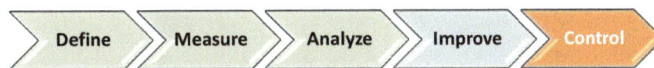

Define 〉 Measure 〉 Analyze 〉 Improve 〉 Control

Control Plans

- Document process changes through SOPs
- Monitor key inputs and outputs
  - Control charts
  - Check lists
  - Mistake proofing
  - Visual management
- Create a corrective action plan to address anomalies

2

Closing the Project

- Hand control back to the process owners

- Celebrate success

- Share lessons learned with the organization

3

Lean Tools

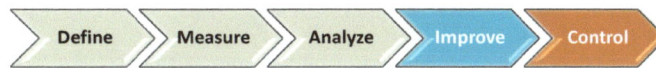

1

Five Lean Principles

01	02	03	04	05
Specify Value	Identify the Value Stream	Create Flow	Pull from the Customer	Seek Perfection
•Value as seen from the end user's point of view	•Represents all activities from order to payment after delivery	•Product moves through process steps without stopping	•Produce only when replenishment is needed	•Requires continuous improvement

2

The Lean Approach

- Focus on eliminating non-value added activities

Transportation	Inventory	Motion
Waiting	Overproduction	Overprocessing
Defects	Skills	8 Wastes

3

Lean Features

- Process focus

- Team focus

- Hands-on focus

- Continuous improvement focus

4

Lean Results in...

- Faster cycle times
- Faster change over times
- Little to no inventory
- Just-in-time methods
- Pull systems
- Continuous flow
- Production leveling
- High quality levels
- Higher employee morale
- **Reduced costs**
- **Higher customer satisfaction**

The Lean Enterprise is known for speed, flow and quality.

5

Visual Management

Visual Management makes the current status of key inputs, outputs and processes apparent at a glance.

Inexpensive

Simple

Unambiguous

Immediate

6

Goal of Visual Management

Problems are easily detected

↓

so that they can be prevented or minimized

↓

to reduce variation and defects

© 2017 University Training Partners

7

| Standardization | Safe environment | Clean, organized environment |
| Progress and performance immediately visible | No blame assigned | Action occurs at signal |

"Visual Factory" Features

8

Visual Management Tools

- Color coding – floor tape, racks, bins
- Machine indicator lights (Andon)
- Operator call lights
- Work instructions
- Maintenance charts
- SPC Charts
- Kanban bins
- Display boards

9

Color Coding

10

Eight Wastes

- Focus on eliminating non-value added activities

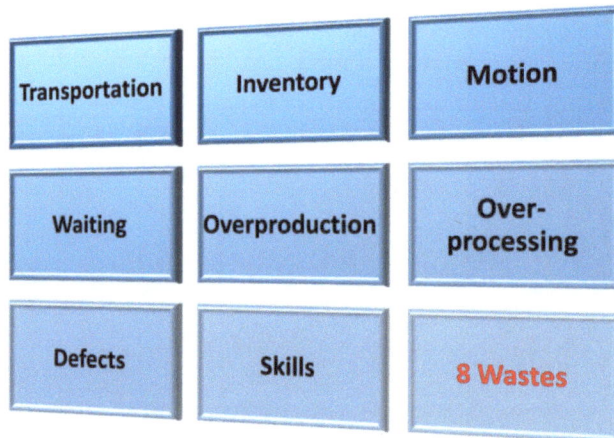

11

The 5S Method

Advantages of 5S

- Productivity (Speed)
- Reduced variation
- Problem identification

Sort
Set in order
Shine
Standardize
Sustain

Mistake Proofing

- Also known as **poka yoke**
- Especially useful in preventing defects when they are rare
- Effective in eliminating random, non-systematic errors

Types of Mistake Proofing

- Control or Warning
  - Machine or process shuts down or signals when an error occurs
  - Lights or sounds alert operator
  - Prevents defects from moving on to the next stage
- Prevention
  - Design does not allow defects from occurring in the first place
  - 100% elimination of defect type

Examples of Mistake Proofing

Diesel and unleaded fuel nozzles	Key release and gear shift	Oil warning light
Marking which leg is to be operated on	Safety seals on packaging	

Lean: Eight Wastes

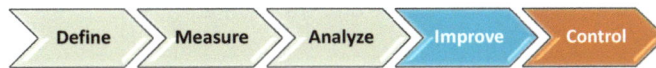

Define	Measure	Analyze	Improve	Control

1

Eight Wastes

- Focus on eliminating non-value added activities

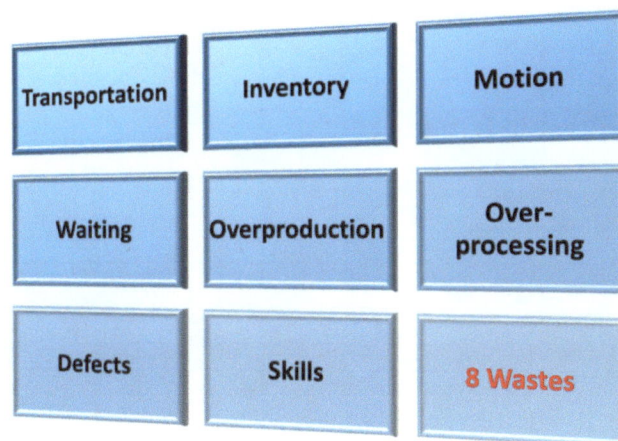

Transportation	Inventory	Motion
Waiting	Overproduction	Over-processing
Defects	Skills	**8 Wastes**

2

Transportation

Definition:
Movement of things:
- Paperwork
- Material
- Electronic information

Risks:
Damage or loss
Delay in availability

3

Inventory

Definition:
Excess:
- Paperwork
- Material
- Supplies
- Equipment

Risks:
Reduced cash flow
Lost productivity due to searching
Excess space
Damage
Obsolescence

Signs:
Stockpiles of supplier, forms, materials
Disorganized storage areas

4

Motion

Definition:
Movement of people

Signs:
Hand carrying
Functional layout
Traveling to shared equipment

Risks:
Reduced capacity
Injury

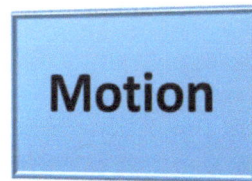

5

Waiting

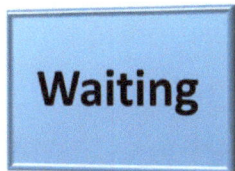

Definition:
People waiting for people,
information, product or
machines

Information, product or
machine waiting for people

Signs:
Idle people
Idle equipment

Risks:
Long lead times
Long work hours
Paid overtime
Capital expenditures for
 equipment

6

Overproduction

Definition:
Producing too much too soon

Risks:
Long lead times
Increased complexity

Overproduction

Signs:
Build up of
work-in-process (WIP)

Build up of queues of material
or people

Over-processing

Definition:
Doing more than the
customer is willing to pay for

Risks:
Long lead times
Low productivity
Frustrated workforce

Signs:
Inspections & audits
Redundant tasks
Too many handoffs

Over-processing

8

Defects

Defects

Definition:
Internal or external suppliers provide incorrect, incomplete or late information or material

Signs:
Correcting material or information that has been supplied
Adding missing information
Clarifying information

Risks:
Rework
Scrap
Dissatisfied customers
Long lead times
Warranty costs

oops!

Skills

Definition:
Not using people's knowledge, skills, aptitude or creativity

Skills

Risks:
Frustrated workforce
Absenteeism
Turnover

Signs:
Excessive reviews or approvals
Specialized workers
Processes designed by managers
Excessive hand-offs

Root Causes of Eight Wastes

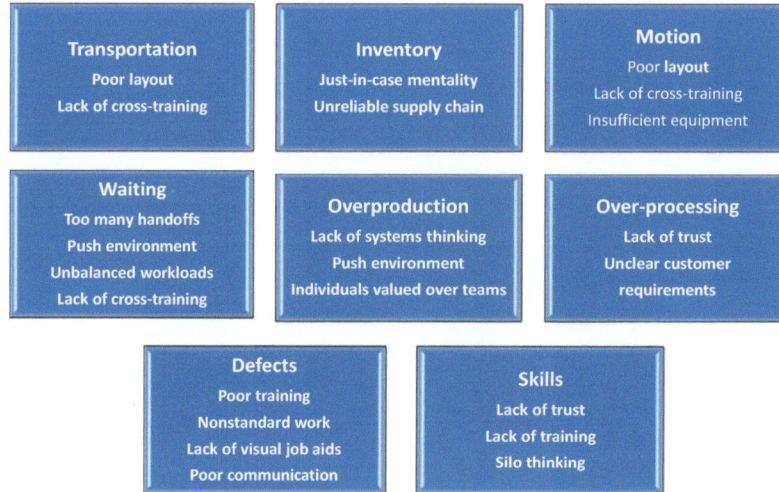

Transportation
Poor layout
Lack of cross-training

Inventory
Just-in-case mentality
Unreliable supply chain

Motion
Poor layout
Lack of cross-training
Insufficient equipment

Waiting
Too many handoffs
Push environment
Unbalanced workloads
Lack of cross-training

Overproduction
Lack of systems thinking
Push environment
Individuals valued over teams

Over-processing
Lack of trust
Unclear customer requirements

Defects
Poor training
Nonstandard work
Lack of visual job aids
Poor communication

Skills
Lack of trust
Lack of training
Silo thinking

11

Lean: Changeover Reduction

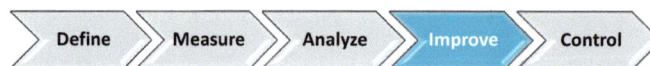

Define Measure Analyze **Improve** Control

12

Changeover Reduction

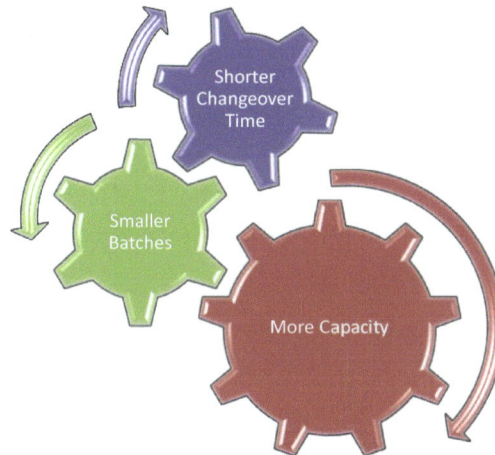

Changeover Time

- **Changeover time** is the total time from the **last unit of production** to the **first unit of good production at full speed**

| Last unit of production | → | CHANGEOVER | → | First unit of good production, full speed |

SMED

- Dr. Shigeo Shingo developed the **"Single Minute Exchange of Dies"** (SMED) concept

- Changeover was reduced from hours to under ten minutes ("single minute")

- Convert internal work to external work

15

SMED Approach

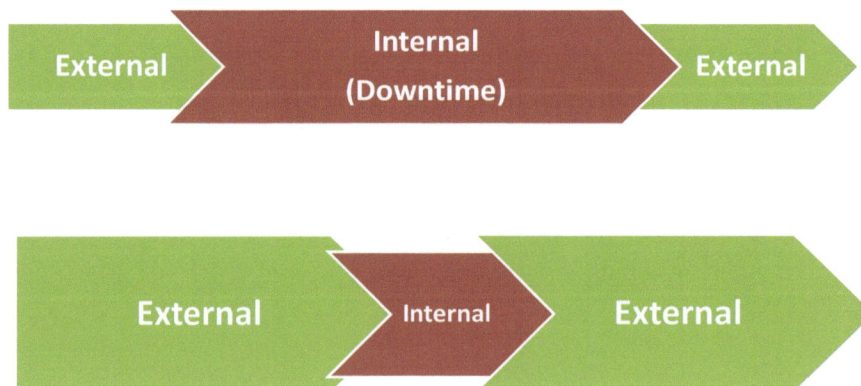

16

Formula 1 Pit Crew

https://www.youtube.com/watch?v=LOJbM0aXZp0

17

SMED Methods

- Staging carts
- Transport fixtures
- Alignment guides
- Quick fasteners
- Interchangeable parts
- Work breakdown and allocation
- Practice, practice, practice

Redesign, Reengineer, Retool

18

Mirasol Project Reports

Jim Sutton – Project Lead

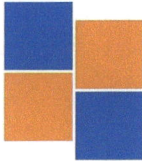

AC Complaints Project

Six Sigma Project Summary					
Project Title	**Scope (location, area, line, department, process boundaries)**				
Reduce AC Complaints	Maintenance and housekeeping departments, contractor				
Process Name	**Start Date**			**Expected End Date**	
AC system	3/15/20xx			6/1/20xx	
Problem Statement	**Expected Benefits**				
Over the last six months of 2011, AC complaints have increased to 3.3% of turnovers resulting in unhappy guests and room change requests.	Reducing AC complaints will increase customer satisfaction and improve percentage of returning guests. Eliminating AC complaints will save $3500 in comped meals & spa treatments, and room upgrades per year.				
Goal Statement	**Key Metrics**				
By June, address AC problems and eliminate AC complaints.	Measure	Units	Baseline	Goal	Goal Met?
	Complaints	Count/ turnvoer	3.30%	0%	Yes
	Temp	F	73	67	68/No
Team Members					
Name	**Role**	**Name**		**Role**	
Jim Sutton	Lead	Joe McGill		Member	
Jon Sand	Champion				
Ed Durst	Member				

AC Complaints Project

Measles Chart	Root Causes Identified
	Ducting on south end of floor 2 is poorly installed PM schedule inadequate Southern facing windows heat rooms Thermostats not set low enough during the day in summer, AC can't catch up

Improvement Ideas	Pilot Study Results
Install new ducting Work with vendor on PM schedule Install film on southern facing windows Have housekeeping turn down temp in southern facing rooms to 67 on days with expected temp > 90	Ducting reinstalled instead of replaced New PM schedule created/techs retrained Canopy installed instead of film, new plantings Worked with Housekeeping on new thermostat procedure

---

AC Complaints Project

Verification of Benefits	Lessons Learned
Southern facing room temps have been holding at 68 degrees on hottest days No AC complaints in May or June	Mechanical problems don't necessarily have to cost a lot of $ to fix Repair before replacing Many little improvements can add up The measles chart was the key in finding the causes of the problem

Control Plan

Key Metrics	Target	How Monitored	Responsible Party	Signatures
AC complaints/turnover	0	Weekly dashboard	Jim Sutton	Champion _____
Room temps	67	X-bar R chart	Joe McGill	Leader _____
Housekeeping % compliance	100	Check sheet	Nina Sanchez	Process Owner _____
PM % compliance	100	Check sheet	Ed Durst	**Close Date** 7/1/20xx

Golf Cart Availability Project

Project Title		Scope (location, area, line, department, process boundaries)				
Increase golf cart availability		Engineering department				
Process Name		**Start Date**		**Expected End Date**		
Engineering garage		3/15/20xx		7/15/20xx		
Problem Statement		**Expected Benefits**				
During June - August of 20xx, there were a total of 12 complaints about late shuttles (1.7% of turnovers), resulting in unhappy guests and comped meals and spa treatments.		Inceasing golf cart availability wil lincrease customer satisfaction and reduce customer complaints about island transportation. Solving the availability problem will save $1000 in comped meals and spa treatments.				
Goal Statement		**Key Metrics**				
Increase golf cart availability to 95% and elimiante shuttle complaints within 4 months.		Measure	Units	Baseline	Goal	Goal Met?
			Count/ changeover			
		Complaints	changeover	1.70%	0	
		Availability	%	80	95	Yes
Team Members						
Name	**Role**	**Name**			**Role**	
Jim Sutton	Lead/YB	Ron Ridpath			Member	
Jon Sand	Champion/GB	Tully Franklin			Member	
Sam Napier	Member					

Golf Cart Availability Project

5S Before and After Photos	Root Causes Identified
	Not enough room for required parts
	Supplier is unreliable/slow
	PM schedule not correct
	Carts are old
	Disorganized inventory system
Improvement Ideas	**Pilot Study/ Improvement Actions**
Perform a 5S on the cart garage	5S created 10% free space, found $2500 in "missing" replacement parts
Identify new parts suppliers	Two new suppliers identified, one under contract
Rework PM schedule	PM rotation schedule reworked, and new assignment board created
Develop a cart replacement schedule	Cart replacement schedule
Computerize parts inventory	Bar code and computer system deemed too much for our small shop; worked with intern to develop an Access database system inhouse

Golf Cart Availability Project

Verification of Benefits	Lessons Learned
Parts now at recommended levels	Keeping an organized garage is key
One cart taken out of service, replacement ordered	Don't be afraid to find new vendors if needed
Carts now operating at 98% availability	Buying your way out of a problem might not be necessary
No complaints for last 2 months	Lean tools can provide quick and effective improvements
	A Six Sigma project doesn't have to be intimidating
	Remember to take before and after pictures when doing a 5S

Control Plan

Key Metrics	Target	How Monitored	Responsible Party	Signatures	
% PM compliance	100	Weekly dashboard	Ron Ridpath	Champion	_____
Parts inventory	Varies by part	Weekly database report	Sam Napier	Leader	_____
Supplier on-time delivery	100%	Monthly report	Tully Franklin	Process Owner	_____
Availability	100%	P chart	Ron Ridpath		
Complaints	0%	Weekly report	Jim Sutton	Close Date	

Mirasol Project Reports

Eli Guzman – Project Lead

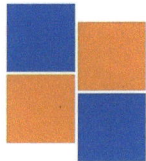

UNIVERSITY TRAINING
PARTNERS

Laundry Water Project

Six Sigma Project Summary					
Project Title	**Scope (location, area, line, department, process boundaries)**				
Decrease laundry water usage	Housekeeping department, laundry team, contractor				
Process Name	**Start Date**		**Expected End Date**		
Commercial laundry facility	1/15/20xx		9/15/20xx		
Problem Statement	**Expected Benefits**				
Water usage in the commecial laundry facility has increased by 25% year over year, resulting in an additional $15,000 in expenses.	Savings due to reduction in water usage to previous levels: Cost to repair/replace system: XXXXX First year net $ savings: XXXXX Subsequent years net savings: XXXXXX				
Goal Statement	**Key Metrics**				
Reduce laundry water usage to previous levels within 8 months.	Measure	Units	Baseline	Goal	Goal Met?
	Efficiency	Gal/lb	5.0	4.0	2.5/Yes
	Rework	%	5	5	Yes
	Cycle time	Min/lb	0.5	0.5	0.48/Yes
	Cpk	NA	0.111	1.67	1.55/No
Team Members					
Name	**Role**	**Name**		**Role**	
Eli Guzman	Lead	Joe Ballard		Member	
Fred Sand	Champion				
Veronica LeMay	Member				

Laundry Water Project

Histogram	Root Causes Identified
Histogram of Water Usage Per Pound	Workers adjusting settings without documenting changes, causing increased variability in efficiency There may be a leak in the main water line Machines inefficient Machine settings incorrect

Improvement Ideas	Pilot Study/ Improvement Results
Have contractor recertify settings, bi-annually after that Install waste water recycling system Install new machines Have water lines inspected for leaks Control who adjusts machines	Contractor reset machines Waste water recycling sysetm deemed too expensive Switched to new high efficiency washer-extractors with built-in rinse water recycling Small leak found in water line; county responsible for repair Trained line workers on proper machine settings

---

Laundry Water Project

Verification of Benefits	Lessons Learned
County has repaired leak in water line New machines have been running at 2.5 gal/lb, std dev = 0.1 gal Rework percentage has not increased. New machines have not increased cycle time July net savings: 35% ($ figure redacted) Expected annual net savings: 40% ($ figure redacted)	The entire department was involved in planning the washer install which made the transition run very smoothly Laundry line workers are now responsible for monitoring machine settings Workers are now trained to set machines themselves

Control Plan				Signatures	
Key Metrics	**Target**	**How Monitored**	**Responsible Party**	Champion	_____
Efficency	2.5 gal/lb	X-MR chart	Joe Ballard	Leader	_____
Rework	10%	P Chart	Veronica LeMay	Process Owner	_____
Machine settings	Varies	Daily check sheet	Line workers		
Cpk	1.67	Via X-MR chart	Eli Guzman	**Close Date**	9/7/20xx

Rooms Not Ready Project

Six Sigma Project Summary					
Project Title		**Scope (location, area, line, department, process boundaries)**			
Reduce complaints for "rooms not available at check in"		Housekeeping and desk			
Process Name		**Start Date**		**Expected End Date**	
Housekeeping room turnover		4/15/20xx		7/15/20xx	
Problem Statement		**Expected Benefits**			
In the past year, Mirasol has received 36 complaints about rooms not being ready at check-in (with a mean shift in July) resulting in unhappy guests.		Increased turnover efficiency will reduce customer complaints at check in and save $2000 annually in comped meals/spa treatments while new guests wait. Increase efficiency will result in a cost avoidance of hiring a new housekeeper in high season at $11/hour.			
Goal Statement		**Key Metrics**			
Eliminate "room not ready at check-in" complaints within 2 months.		Measure	Units	Baseline	Goal	Goal Met?

Measure	Units	Baseline	Goal	Goal Met?
Room not ready at check-in complaints	Count/turnover	3.50%	0%	Yes
Room cleanliness complaints	Count/turnover	0	0	Yes

Team Members			
Name	**Role**	**Name**	**Role**
Eli Guzman	Lead	Nina Sanchez	Member
Fred Sand	Champion	Cindy Sparks	Member
Dave Harris	Member		

Rooms Not Ready Project

Histogram

Histogram of Turnover time

Root Causes Identified

Delays are caused by housekeepers having to dump trash and walk laundry to facility during turnover
Carts not loaded with needed supplies - housekeepers walking to supply closets on first floor
Duvet covers have 25 buttons that must be checked and rebuttoned
Special pillow requests are not communicated in time

Improvement Ideas

Preload carts with correct supplies the night before turnover days
Form a "turnover team" using SMED techniques
Create detailed process flow for efficient room cleaning
Print pillow and other special requests te night before, load on cart
Buy new duvets with zippers

Pilot Study Results

Room turnover reduced to an average of 29 minutes, std dev = 45 seconds
Team floater dumps trash and takes laundry to facility
New duvet covers are much easier to handle
All rooms turned over within the 3 hour window with current staffing levels
Time to refresh rooms also reduced to 14 minutes, an extra bonus

Rooms Not Ready Project

Verification of Benefits				Lessons Learned			
New system used on three high turnover Saturdays in the high season - all rooms ready for guests by check in				Tools used: spaghetti diagram, process flow chart, SMED			
No additioanl staff required				The housekeeping staff was involved in developing new procedures and			
No reduction in room cleanliness				brainstormed ways to reduce errors. This made the group accept the changes			
All housekeepers trained on new procedures				in their workflow, and improved housekeeping morale.			
Attendance increased from 85% to 90% on turnover days				Housekeeping attendance has improved, but more needs to be done			

Control Plan				Signatures	
Key Metrics	**Target**	**How Monitored**	**Responsible Party**	Champion	_____
Turnaround time	29 min	EWMA chart	Eli Guzman	Leader	_____
Attendance	100%	P Chart	Nina Sanchez	Process Owner	_____
Room complaints	0	Run chart	Eli Guzman		
				Close Date	7/28/20xx

www.ingramcontent.com/pod-product-compliance
Lightning Source LLC
Chambersburg PA
CBHW041442210326
41599CB00004B/105